Risk-based, Management-led, Audit-driven,

SAFETY MANAGEMENT SYSTEMS

Risk-based, Management-led, Audit-driven,

SAFETY MANAGEMENT SYSTEMS

RON C. McKINNON

CRC Press
Taylor & Francis Group
Boca Raton London New York

CRC Press is an imprint of the
Taylor & Francis Group, an **informa** business

CRC Press
Taylor & Francis Group
6000 Broken Sound Parkway NW, Suite 300
Boca Raton, FL 33487-2742

© 2017 by Taylor & Francis Group, LLC
CRC Press is an imprint of Taylor & Francis Group, an Informa business

No claim to original U.S. Government works

Printed on acid-free paper
Version Date: 20161019

International Standard Book Number-13: 978-1-4987-6792-7 (Hardback)

Visit the Taylor & Francis Web site at
http://www.taylorandfrancis.com

and the CRC Press Web site at
http://www.crcpress.com

Contents

List of Figures

Preface

This book explains how accidents are caused and how they can be prevented by the implementation of a safety management system (SMS) that is based on risks arising from the business, which is initiated and led by management at all levels, and which is constantly monitored by audit processes. It explains how accidental losses are caused, and how a safety system reduces the probability of them occurring.

Practical and basic risk assessment techniques are discussed, as well as the authorities, responsibilities, and accountabilities management needs to assign to make a safety system function successfully. The safety system audit process is explained in simple terms, and its importance in continual improvement is highlighted.

Using an example safety management system (Example SMS) comprising 5 sections and 84 elements, this book explains each element's role in the system in detail. Examples of standards are given, showing the link between safety theory and practice. This book shows how management principles are translated into practical actions at the workplace.

An introduction is given to national and international Guidelines for safety and health management systems, and the Example SMS used in this book shows how to comply with the requirements of these Guidelines.

While traditional injury rates and downstream measurements tell us little about the state of safety at an organization, an SMS is a proactive set of processes that are management performances that can constantly be measured.

Acknowledgments

This publication is based on many years of work implementing safety management systems (SMSs) at many organizations, in different countries. It is also based on nearly 40 years of advising and guiding others on how to implement an SMS at their organizations. Much of the knowledge and information in this book was gained from the people that I have associated with, and worked with in the safety profession. I thank them and my mentors for sharing their knowledge and experience with me. It was a pleasure knowing and working with you. I pay tribute to the safety pioneers that I have quoted in this book. They need to be thanked for their diligent research into safety management, and especially the pioneering of SMSs.

For making this publication possible I thank my wife, Maureen McKinnon, who spent numerous weeks editing this manuscript. This support warrants my deep gratitude.

Note: Every effort has been made to trace rights holders of quoted passages and researched material, but if any have been inadvertently overlooked, the publishers would be pleased to make the necessary arrangements at the first opportunity.

About this Book

Risk-based, Management-led, Audit-driven, Safety Management Systems explains what a safety management system (SMS) is, and how it reduces risk in order to prevent accidental losses in an organization. It advocates the integration of safety and health into the day-to-day management of the enterprise as a value, rather than an add-on.

This book refers to international Guidelines on SMSs, as well as the proposed International Organization for Standardization (ISO) 45001, which could soon become the international safety benchmark for organizations worldwide.

Emphasis is based on the identification and control of risk as the basis for an SMS. Examples of a simple risk matrix and daily task risk assessment are given in this book, as well as a simplified method of assessing, analyzing, and controlling risks.

In no mean terms, this book emphasizes that the safety movement must be initiated, led, and maintained by management at all levels. The concepts of safety authority, responsibility, and accountability are described as being the key ingredients to safety system success. Safety system audits are expounded in simple terms, and leading safety performance indicators are suggested as the most important measurements in preference to lagging indicators.

An example, 5-section, 84-element SMS, is used throughout the book to explain the components and elements of an SMS. Each of the 84 elements is described in detail as to how they dovetail into the system. Risk ranking of elements, dependent on risk or benefit, is also simplified.

Using simple, understandable examples, the chapters give a complete overview of an SMS and its components. The Example SMS used in this book conforms to most of the SMS Guidelines published by leading world authorities, and enables an organization to structure its own world class SMS, based on this example.

About the Author

Ron C. McKinnon, CSP (1999–2016), is an internationally experienced and acknowledged safety professional, author, motivator, and presenter. He has been extensively involved in safety research concerning the cause, effect, and control of accidental loss, near miss incident reporting, accident investigation, safety promotion, and the implementation of safety management systems for the last 40 years.

Ron McKinnon received a national diploma in technical teaching from the Pretoria College for Advanced Technical Education, a diploma in safety management from the Technikon SA, South Africa, and a management development diploma (MDP) from the University of South Africa, Pretoria. He received a master's degree in safety and health engineering from the Columbia Southern University, Alabama.

From 1973 to 1994, Ron McKinnon worked at the National Occupational Safety Association of South Africa (NOSA), Pretoria, South Africa, in various capacities, including general manager of operations and then marketing. He is experienced in the implementation of safety systems, auditing safety systems, and safety culture change interventions. During his tenure with NOSA, he implemented safety systems and conducted training in numerous countries.

From 1995 to 1999, Ron McKinnon was safety consultant and safety advisor to Magma Copper and BHP Copper North America, respectively. At BHP Copper he was a catalyst in the safety revolution in the copper industry that resulted in an 82% reduction in the injury rate, and an 80% reduction in the severity rate.

In 2001, he spent two years in Zambia introducing world's best safety practices to the copper mining industry. Thereafter he accepted a two-year contract in the Kingdom of Bahrain, Arabian Gulf, where he successfully facilitated a safety culture change at the country's second largest employer.

After spending two years in Hawaii at the Gemini Observatory, he returned to South Africa. He recently contracted as the principal consultant to Saudi Electricity Company (SEC), Riyadh, to implement a world's best practice safety management system (Aligned to OHSAS 18,001), throughout its operations across the Kingdom involving 33,000 employees, 27,000 contractors, 9 consultants, and 70 Safety Engineers.

Ron C. McKinnon is the author of *Cause, Effect and Control of Accidental Loss,* published by CRC Press/Taylor & Francis Group in 2000. He is also the author of *Changing Safety's Paradigms,* published in 2007 by Government Institutes (USA), as well as *Safety Management, Near Miss Identification, Recognition and Investigation,* published by CRC Press/Taylor & Francis Group in February 2012. In 2014, *Changing the Workplace Safety Culture* was also published by CRC Press/Taylor & Francis Group.

Ron McKinnon is a retired professional member of the ASSE (American Society of Safety Engineers), Tucson Chapter Past President, and an honorary member of the Institute of Safety Management. He is currently a safety consultant, safety culture change agent, motivator, and trainer, is often a keynote speaker at international safety conferences, and currently consults to international organizations.

1 Introduction

EXTENT OF THE PROBLEM

According to the International Labor Organization (ILO), statistics published in 2015, approximately 2.3 million people died as a result of work-related accidents or diseases (ill health) in 2013. Also, according to the ILO, some 600,000 lives would be saved every year if available safety practices and appropriate information were used.

They quote:

- Every year, 250 million accidents occur causing absence from work, the equivalent of 685,000 accidents every day, 475 every minute, 8 every second.
- Working children suffer 12 million occupational accidents and an estimated 12,000 of them are fatal.
- 3000 people are killed by work every day, 2 every minute.
- Asbestos alone kills more than 100,000 workers every year (ILO website, 2016a).

The National Safety Council's (USA) publication, *Injury Facts* (2013), lists unintentional-injury-related deaths for the year 2011 at 3,905, and medically consulted work injuries at 5,000,000. The total cost of unintentional injuries is given as $753 billion and the comprehensive loss to the U.S. economy is given as $4,364.5 billion for 2011. Work injuries alone cost $188 billion for the same year (NSC, *Injury Facts*, 2013, p. 8).

These are shocking statistics and a heavy burden for society and the economy. Implementing a strong occupational health and safety management system (SMS) helps organizations reduce accidents and ill health, avoid costly prosecutions, perhaps even reduce insurance costs, and create a positive culture in the organization when its people see that their needs are being taken into account.

SAFETY

Safety is the control of all forms of accidental loss by identifying, analyzing, and reducing risks. The main areas of loss which are prevented or reduced by a safety management system (SMS) are as follows:

- Injuries and fatalities to persons
- Occupational diseases and illnesses
- Damage to equipment and property
- Harm to the environment
- Hidden losses such as poor quality, company reputation, etc.

DEFINITIONS

Work Injury

A work injury is any injury suffered by a person, and which arises out of, and during the course of, his or her normal employment. The definition of work injury includes work related disability, occupational disease, and occupational illness.

Occupational Disease

An occupational disease is a disease caused by environmental factors, the exposure to which is peculiar to a particular process, trade, or occupation, and to which an employee is not normally subjected, or exposed to, outside of, or away from, his or her normal place of employment.

Property Damage

Property damage is accidental or unintentional damage or spoilage to equipment, structures, material, or products, caused by an accident or undesired event.

While these type of accidents may seem insignificant, modern safety thinking is that business errors that cause injury and disease have the same symptoms as events that cause damage. The exchange of energy which caused the property or equipment damage could have, under slightly different circumstances, caused injury to persons.

Property Damage Is an Accident

Property damage accidents, therefore, should receive the same attention as injury-producing accidents to identify and rectify the failure in the system. Some of the international safety management system Guidelines discussed in this publication do not include property damage in their recommendations, as they state it does not form part of the safety, health, and welfare protection of employees. So what happens if a cargo container falls and lands near a group of workers? There is no injury, fatality, or illness as a consequence, so according to some Guidelines this is not a concern for the safety management system. Yet, if the container happened to have fallen a few feet to the left or right, there would have been serious injury to one or more of the employees who were under the container. The difference between the outcomes of the same accidental event is fortuitous, the event is what should be investigated irrespective of the consequent.

A safety management system should consider that property damage accidents are accidents which should have been prevented and which, under slightly different circumstances, could have caused injury, fatality, or illness. Because no fatality, injury, or illness took place, is only fortuitous. The root cause of the system failure (the accident) is what needs to be investigated and treated, to prevent a recurrence, irrespective of the outcome.

Areas of Loss

The main areas of loss are to people in the form of death, injury, permanent disability, or disfigurement along with the loss of earning power and, in some cases, quality of life. Another area of accidental loss is damage to equipment, machinery, and product caused by accidents. These losses are merely the tip of the iceberg.

The hidden layer is the indirect costs of these losses which are not compensated or covered by insurance, but which still cost the organization time and resources. The totally hidden costs of accidents are difficult to quantify financially. They include losses such as employee morale, company reputation, legal litigation, fines, etc.

OCCUPATIONAL HYGIENE

DEFINITION

Most sources, including IPM Safety, define occupational hygiene *as the science and art devoted to the anticipation, recognition, identification, evaluation, and control of environmental stresses arising out of a workplace, which may cause illness, impaired well-being, discomfort and inefficiency of employees or members of the surrounding community* (IPM Safety website). Occupational hygiene is also described as the science dealing with the influence of the work environment on the health of employees.

OBJECTIVES OF OCCUPATIONAL HYGIENE

The objective of occupational hygiene is to recognize occupational health hazards, evaluate the severity of these hazards, and to eliminate them by instituting control measures. Some stresses include chemical hazards, exposure to noise, to airborne contaminates, ergonomic stresses, etc. As with any exposure, excessive exposure to any one, or combination of the above agencies could result in occupational disease, injury, or other adverse symptoms. Where the occupational health hazard cannot be eliminated entirely, occupational hygiene control methods must aim to reduce the exposure to the hazard and institute measures to reduce the hazard.

ENGINEERING REVISION

The easiest form of safety and health control is engineering revision, where the equipment is modified and the process is completely contained, suppressed, ventilated, or reduced. This does not always work and may prove to be too costly.

Limiting the exposure of workers to the hazard is also an acceptable control measure, but this may reduce the production and may also prove too expensive. Providing personal protective equipment such as respirators, earmuffs, etc., is a method of control, but is perhaps the least effective and should be viewed as a last resort.

ACCIDENT CAUSATION

Accidents and their consequences can be prevented and the resultant losses spared if enough effort is applied to control workplace risks. Research has shown that less than 2% of undesired events are beyond local control, and these include happenings such as floods, earthquakes, tsunamis, etc. The vast majority of accidents can be prevented by implementing controls, checks, and balances in the form of a structured safety and health management system. Falls of employees to a lower level is a leading cause of work fatalities. Can these be prevented? The answer is yes, they can. Modern technology and fall restraint and fall arrest systems are available and, if applied and enforced by management and worker organizations, can prevent workers from falling to their deaths. The cost of fall protection outweighs the cost of a work fatality and should make good business sense.

SAFETY MANAGEMENT SYSTEMS (SMSs)

Safety and health management systems identify and treat accident causes and not symptoms. To guide management in controlling areas of potential loss, and to set standards, there are existing safety and health management systems that provide excellent system frameworks. These are sometimes referred to as structured safety programs, but the preferable term is *safety management systems*, as they do follow a systems approach and methodology to prevent loss. These systems prescribe certain elements under certain headings and give details of what aspects of a safety management system should be instituted as a foundation for the prevention of accidental loss.

A FORMALIZED APPROACH

A safety and health management system is a formalized approach to health and safety management through use of a framework that aids the identification and control of safety and health risks. Through routine monitoring, an organization checks compliance against its own documented safety and health management system (safety management system), as well as against legislative and regulatory compliances. A well-designed and operated safety management system reduces accident potential and improves the overall management processes of an organization.

RISK-BASED SYSTEM

The safety management system must be a risk-based system. That means it must be aligned to the risks arising out of the workplace. Emphasis on certain safety management system elements will be different according to the hazards associated with the work and the processes used. There is unfortunately no *one size fits all* safety management system that will be ideal for all mines, industries, and other workplaces; therefore they should be seen as a framework on which to build a risk-specific system for the industry. The main aim of any safety system is to reduce risk, therefore the system must be aligned to those risks.

MANAGEMENT-LED SYSTEM

The key factor in safety is management leadership. The safety management system must be initiated, led, and supported by senior management as well as line and front line management.

Safety systems that originate and which are maintained in the safety department will have little effect on the organization. It is estimated that about 15% of a company's problems can be controlled by employees, but 85% can be controlled by management. This means that most safety problems are management problems. Management will also realize that if they can manage the intricate and difficult concept of safety, then they will be able to manage other aspects of management easier, as managing safety enables them to manage more effectively.

AUDIT-DRIVEN SYSTEM

What gets measured usually gets done. Safety is an intangible concept and is traditionally measured after the fact—once a loss has occurred. The safety management system must be an audit-driven system, which calls for ongoing measurements against the standards and quantification of the results.

A safety system converts safety intended actions into proactive activities and assigns responsibility and accountability for those actions, very similar to what a manager does with his or her subordinates. Each activity, usually included in the safety system elements, can then be scored on a 1–5 scale as to whether it has been achieved or not. At the end of the day, by means of audit, the entire system can be quantified by the score allocated. The safety system's effectiveness has been measured. The elements that scored less than full points are highlighted as areas that need improvement.

CONTROL NOT CONSEQUENCE

The following chapter (Chapter 2) analyzes the components of an accident sequence and shows that by identifying the risk at a workplace and implementing a structured, risk-based system, management-led and audit-driven safety management system, these events can be prevented. The accident sequence shows that often the outcome of an undesired event can be swayed by good or bad fortune, and that safety measurements of consequent are not accurate indicators of the safety at an organization. An organization must focus on control and not consequence.

2 Accident Causation

INTRODUCTION

The loss causation domino sequence, originally termed the accident sequence, was originally proposed by H.W Heinrich in 1929, and has been revisited and updated by a number of safety pioneers. Other loss causation models and theories have also been proposed by many safety professionals.

The Cause, Effect and Control of Accidental Loss (CECAL) sequence was proposed by Ron C. McKinnon and published as a book under the same name in 2000. It introduces the three Luck Factors into the accident sequence (McKinnon, 2000).

LOSSES

Losses normally occur as a result of accidents. An accident is defined by Frank Bird Jr. as: *an undesired event, which results in harm to people, damage to property, or an interruption of the work process* (Bird and Germain, 1992, p. 18).

Accidents are caused by a breakdown in the management control system (the safety and health management system), and the end result of every accident is some form of loss. The four main areas of loss are people, equipment, property, and environment.

NEAR MISS INCIDENT

There is confusion as to the differentiation between *accidents* and *incidents*. A near miss incident is defined as: *an undesired event, which, under slightly different circumstances, could have resulted in a loss.* This means that accidents do result in losses, but near miss incidents do not result in any loss. They do, however, offer a warning as to the *potential* of loss occurring.

TRADITIONAL VIEWPOINT

In general, organizations do not normally acknowledge having experienced an accident until there is severe injury or illness to a person or persons. Most undesirable events do not end up in any loss at all. The majority of accidents cause property damage and minor injury, and less than 2% of accidental occurrences result in serious injury. Based on the CECAL theory, the end result of an undesired event is often swayed by fortuity, or Luck Factors, over which an organization has little or no control.

IMPORTANCE

The loss causation analysis is of vital importance to the safety management profession. It calls for a different way of looking at, measuring, and promoting the prevention of occupational injuries, damage, and diseases. The theory clearly demonstrates that traditional forms of safety measurement, and the almost disregard of near miss incidents, has to change before losses, such as the injury toll, can be reduced.

SAFETY MANAGEMENT SYSTEM (SMS)

The CECAL sequence proposed that all forms of accidental loss are triggered by a failure to identify the hazards, to analyze and evaluate the risks, and to institute control measures in the form of a structured and sustained safety management system. This in turn leads to weaknesses in the management system, which gives rise to job and personal factors, commonly referred to as the root causes of accidents. These root causes prompt high-risk acts to be committed and in turn allow unsafe conditions to be created. Once this situation exists, Luck Factor 1 determines whether there will be a contact with a source of energy or not. No contact with a source of energy results in a near miss incident, which is commonly referred to as, "nothing happened."

Should there be contact with a source of energy, Luck Factor 2 then determines the outcome of the exchange of energy. The outcome could be injury, property damage, or business interruption or a combination of two or all three. If the exchange of energy causes personal injury, Luck Factor 3 then determines the severity of the injury. The last domino in the sequence depicts the costs that are incurred as a result of the losses.

COSTS OF ACCIDENTAL LOSS

Examples of recent costs incurred include the BP Deepwater Horizon oil spill. By February 2013, criminal and civil settlements and payments to a trust fund had cost the company $42.2 billion. A U.S. District Court judge ruled that BP was primarily responsible for the oil spill because of its gross negligence and reckless conduct, and in July 2015, BP agreed to pay $18.7 billion in fines, the largest corporate settlement in U.S. history (Fortune.com, 2015).

ACCIDENT SEQUENCE

Accidents are caused by a sequence of events. A series of blunders. A combination of circumstances, and activities, culminate in a loss. The loss may be an injury, damage, or business interruption or a combination thereof.

FAILURE TO ASSESS THE RISK

The accident sequence is triggered by a failure to adequately identify the hazards and assess the risks they pose, which in turn causes a lack of, or inadequate control in the form of a weak or non-existent safety management system. If the risks posed by

FIGURE 2.1 Failure to assess the risk results in a weak safety management system.

the enterprise are not identified and assessed, then controls (in the form of a safety system) are more than likely not in place to prevent the accident sequence from occurring (Figure 2.1).

WEAK OR NON-EXISTENT SAFETY MANAGEMENT SYSTEM

The second link in the accident sequence is a lack of, or inadequate control of risks. This lack of control could be because of no safety system, no safety system standards, non-compliance to the standards, or lack of standards, systems, and controls. This weakness leads to the root causes of accidents (Figure 2.2).

ACCIDENT ROOT CAUSES

The basic or root causes of accidents are categorized as personal and job factors. They are the underlying reasons why high-risk acts are committed, and why unsafe conditions exist. A personal factor could be a lack of skill, physical or mental incapability to carry out the work, poor attitude, or lack of motivation. Job factors could

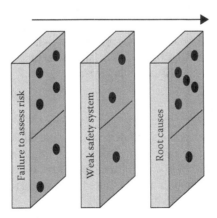

FIGURE 2.2 A weak or inadequate safety system creates root causes of accidents.

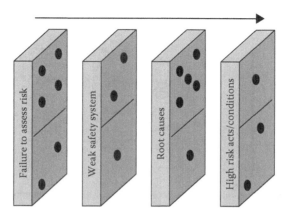

FIGURE 2.3   Root causes lead to high-risk work conditions and acts.

include inadequate procedures, poor maintenance, incorrect tools, or inadequate equipment. The root causes are what trigger immediate accident causes, which are unsafe work conditions and high-risk work practices (Figure 2.3).

UNSAFE (HIGH-RISK) CONDITIONS AND UNSAFE (HIGH-RISK) ACTS

The fourth domino in the accident sequence depicts the high-risk acts and high-risk conditions, commonly referred to as the immediate causes of accidents. A high-risk, or unsafe act, is defined as: *the behavior or activity of a person which deviates from normal accepted safe procedure.* An unsafe condition is defined as: *a hazard or the unsafe mechanical or physical environment.*

UNSAFE (HIGH-RISK) CONDITIONS

Unsafe conditions are physical work conditions, which are below accepted standards, contain hazards, and pose a risk of loss. This results in a high-risk area or an unsafe work environment. The unsafe work conditions such as unguarded machines, cluttered walkways, poor housekeeping, inadequate lighting, poor ventilation, inadequate procedures, etc., are responsible for a large percentage of all accidents and are as a result of a weak or non-existent safety system.

UNSAFE (HIGH-RISK) ACTS

High-risk acts are the behaviors that put people at risk. This means that people are behaving contrary to the accepted safe practice and are thus creating a hazardous, high-risk situation, which could result in accidental loss. In some instances, high-risk acts are committed knowingly, and sometimes they are as a result of poor communication or lack of specific knowledge.

High-risk acts include working without authority, failure to warn somebody, rendering safety devices inoperative, or clowning and fooling around in the workplace.

Numerous accidents are caused by high-risk acts. All high-risk acts and conditions can be traced to their root causes created by inadequate safety system control, as a result of a failure to assess and manage risk. All accidents have multiple causes and an organization should not support certain philosophies that blame accidents entirely on human behavior.

LUCK FACTOR 1

Failure to assess and control the risk results in poor control in the form of an inadequate safety management system, weak standards, or deviations from standards. This creates root causes in the form of personal and job factors. These root causes lead to high-risk behaviors and unsafe conditions. The stage has been set, and invariably the next event is either a contact with the source of energy and a loss, or a near miss incident, depending on the Luck Factor 1 (Figure 2.4).

Many safety pioneers identified the Luck Factors when they stated that there is no logical explanation why some high-risk acts end up as accidents (loss) and the same high-risk act, under slightly different circumstances, ends up as a near miss incident (no loss). The difference in outcomes must be fortuitous, as one cannot control or predict the consequence of a high-risk action or hazardous condition. The difference, therefore, between a near miss incident (close call) and an accident that produces a loss, can only be ascribed to luck; good or bad fortune.

UNDER SLIGHTLY DIFFERENT CIRCUMSTANCES

The cause and effect of accidental loss has shown that failure to assess and control the risks creates root causes, which in turn produces an environment that condones high-risk acts and unsafe conditions. These, as swayed by Luck Factor 1, either result in a contact (exchange of energy), a near miss incident, or nothing happens (Figure 2.5).

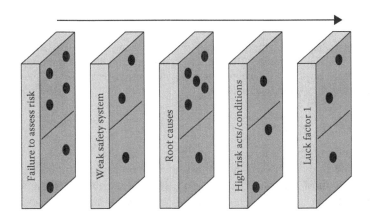

FIGURE 2.4   Luck Factor 1.

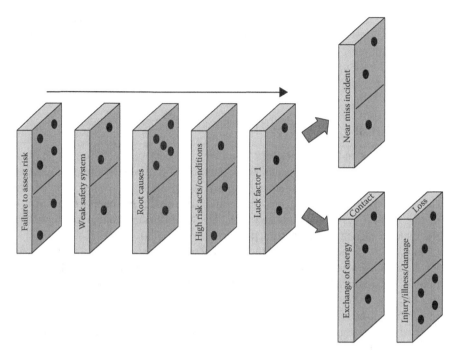

FIGURE 2.5   The Luck Factor resulting in a near miss incident with no loss.

WARNINGS

Near miss incidents have often been called near misses, narrow escapes, and warnings. Irrespective of their terminology, they are the real safety in the shadows, the accidents that the organization hasn't had yet. We have been aware of near miss incidents, near hits, and warnings for a number of years. Safety pioneer Heinrich (1959) introduced his 10 axioms of industrial safety back in the 1930s. His third axiom reads as follows:

> The person who suffers a disabling injury caused by an unsafe act, in the average case, had over 300 narrow escapes from serious injury as a result of committing the very same unsafe act. Likewise, people are exposed to mechanical hazards hundreds of times before they suffer injury. (p. 21)

POTENTIAL FOR LOSS

The sequence of events described so far have encountered Luck Factor 1 and have resulted in either a contact with a source of energy or a near miss incident. Traditionally, most near miss incidents are ignored because "nothing happened" or so it seems. In referring to numerous safety pioneers' accident ratios, there are numerous near miss incidents that occur for every serious injury experienced. Every serious injury has been preceded by numerous warnings in the form of near miss

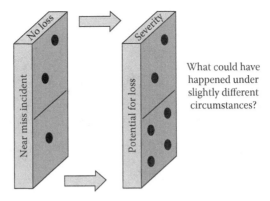

FIGURE 2.6   Near miss incidents have potential for loss.

FIGURE 2.7   The iceberg effect with the serious injury visible above the waterline.

incidents, close calls, high-risk actions, and other hazardous situations. As in the iceberg effect, the injury is what most organizations see and react to. Many close calls, near miss incidents, and other deviations occur, but remain hidden below the waterline. Since many have potential to cause serious loss, under slightly different circumstances, it would then seem logic that near miss incidents should be treated as seriously as loss producing accidents (Figure 2.6). Deep beneath these events lie high-risk acts and unsafe conditions (Figure 2.7).

EXCHANGE OF ENERGY

Once a high-risk act has been committed, or an unsafe condition exists, Luck Factor 1 determines whether there will be a contact and exchange of energy or not. This exchange of energy is a contact with a substance or source of energy greater than the threshold limit of the body or article (Figure 2.8). The high-risk act or unsafe condition may result in a near miss incident, where, although there was no contact and exchange of energy, there was potential to cause loss.

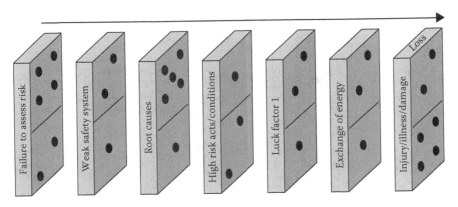

FIGURE 2.8   An exchange of energy and a contact results in a loss.

EXCHANGE OF ENERGY BUT NO LOSS

Sometimes there could be a flow of energy, but no contact and no loss. In some instances there is contact but the exchange of energy is lessened by, for example, the use of personal protective equipment, so there is no loss. When all other means to isolate or protect from the source of energy have failed, personal protective equipment (PPE) is worn to lessen and protect against the flow of energy from accidental contacts. The wearing of PPE is regarded as the last resort for safeguarding, as permanent methods of hazard elimination are better long-term solutions. In some events, the energy exchanged is below the threshold level and does no harm. The contact, or exchange of energy, is the part of the sequence that injures, damages, pollutes, or interrupts the business process.

LUCK FACTOR 2

The failure to assess risks and institute the necessary controls creates accident root causes. They in turn lead to the high-risk act and/or unsafe condition, which, because of Luck Factor 1, results either in a near miss incident (no loss) or a contact and exchange of energy (loss).

TYPES OF LOSS

Luck Factor 2 also sways what type of loss could be caused by an accidental exchange and contact with a source of energy (Figure 2.9). Examples of loss types are injury or disease, property and equipment damage, business interruption or delay, or a combination thereof.

Traditionally, most safety management system efforts focused on one specific type of loss—injury to people, which in turn is influenced by luck factors. Many situations that could have killed or injured workers never resulted in injury or death and were merely a case of good fortune, not good safety. A good safety management system should also focus on the non-loss causing outcomes of undesired events. It should emphasize *precontact* control rather than *post-contact* control. It must direct its energy to prevent the event from occurring rather than depending on the consequence.

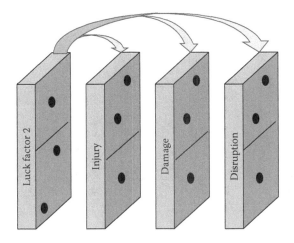

FIGURE 2.9   Luck Factor 2.

PROPERTY AND EQUIPMENT DAMAGE AND BUSINESS INTERRUPTION

The exchange of energy caused by a high-risk act, unsafe condition, or combination thereof, could result in injury, damage to property and equipment, or business interruption as determined by Luck Factor 2 (Figure 2.10).

LUCK FACTOR 3

Heinrich (1959) compiled 10 axioms of industrial safety, the most pertinent one to this chapter being axiom 4 that states "the *severity* of an injury is largely fortuitous—*the occurrence* of the accident that results in injury is largely preventable" (p. 21).

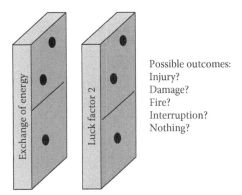

FIGURE 2.10   Dependent on Luck Factor 2, the outcome could be injury, damage, disruption, or a combination of two or all three.

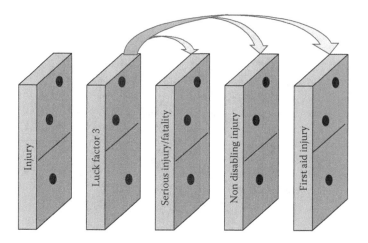

FIGURE 2.11   The severity of an injury is determined by Luck Factor 3.

This fourth axiom is perhaps the most significant statement in the safety management profession. What Heinrich states is that the degree of injury depends on luck but that the accident can be prevented. What he further indicates by this axiom is that while the accident (the event) can be prevented, the severity is something over which we have little or no control (Figure 2.11).

COST OF ACCIDENTAL LOSS

All accidents cost money (Figure 2.12). Whether the outcome of an accident is a fatality, injury, property damage, or a business interruption, financial losses are incurred. Most costs after an accidental loss are hidden. Many are difficult to calculate but are losses nevertheless. Examples of totally hidden costs could be the organization's reputation, loss of customers, employee morale, etc (Figure 2.13).

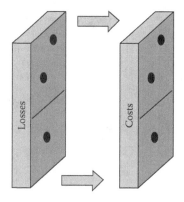

FIGURE 2.12   Every accidental loss incurs costs to the organization.

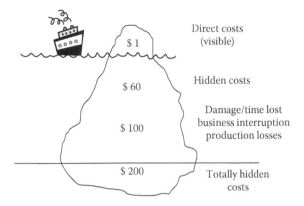

FIGURE 2.13 Visible and hidden accident costs.

SAFETY MANAGEMENT SYSTEM

The loss causation analysis of an accident is of vital importance to the safety management profession. It calls for a different way of looking at, measuring, and controlling the prevention of occupational injuries, damage, and disease. The sequence clearly demonstrates the need for a structured safety management system to identify the business risk and to institute ongoing management controls to prevent the sequence occurring. The safety management system prevents the accident domino effect by constantly identifying risks and setting up controls, balances, and checks that eliminate root accident causes. This stabilizes the control domino preventing it from toppling and creating the domino effect.

CONCLUSION

Only once the risks have been identified and controls in the form of a structured safety management system are put in place, will accident root causes and high-risk acts and conditions be reduced along with consequent loss (Figure 2.14).

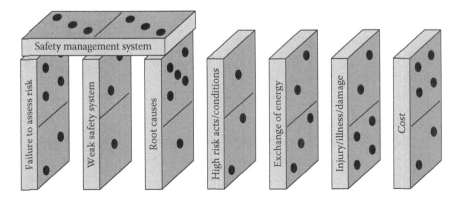

FIGURE 2.14 A safety management system stabilizes the factors that lead to accidental loss.

3 Safety Management Systems and Guidelines

INTRODUCTION

A safety and health management system (SMS) (safety system) is a formalized approach to health and safety management through use of a framework that aids the identification, control, and mitigation of safety and health risks. Through routine monitoring, an organization checks compliance against its own documented safety and health management system (SMS), as well as legislative and regulatory compliance. It is a series of ongoing management processes. A well-designed and operated safety management system reduces accidental loss potential and improves the overall management processes of an organization. Introducing a formalized SMS is the only way to change an organization's safety culture.

A SYSTEMS APPROACH TO SAFETY

A management system is the framework of policies, processes and procedures used to ensure that an organization can fulfill all tasks required to achieve its objectives.

A basic system is based on the *Plan, Do, Check, Act* methodology (Figure 3.1). A complete safety system would include assignment of personal authority, responsibility and accountability, and a schedule for activities to be completed. Part of a systems approach are the auditing tools to identify and implement corrective actions and thus create a process of sustained continuous improvement.

THE PLAN, DO, CHECK, ACT METHODOLOGY

Another process approach, based on the *Plan, Do, Check, Act* methodology, has six steps starting with the safety and health policy, the planning, the implementation and operation, and after that the checking and necessary corrective action. The last step is the management review of the progress and this cycle leads to a continual safety improvement process. (Figure 3.2)

ISSMEC

The ISSMEC, *Identify, Set Standards of Accountability and Measurement, Measure, Evaluate, and Correct* management technique is also often used as a framework for safety management systems (Figure 3.3). Setting standards refers to standards of measurement as well as standards of authority, responsibility, and accountability. Corrective actions are called for once deviations are highlighted and commendation is given for compliance to standards.

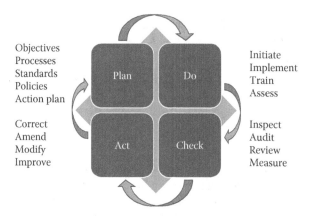

FIGURE 3.1   The Plan, Do, Check, Act (PDCA) methodology.

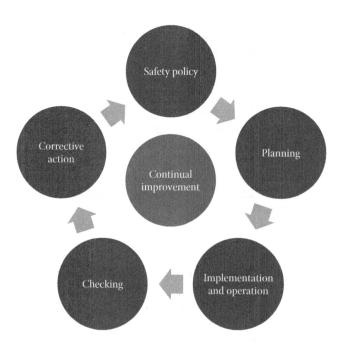

FIGURE 3.2   Another example of a process approach to safety management.

ONGOING PROCESS

A safety management system is a continuous, ongoing process which enables an organization to control its occupational health and safety risks and to improve its safety and health endeavors by means of continuous improvement of safety and health processes. The achievement of targets and goals must be sustained year in and year out. An organization will never be able to state that they have completed the

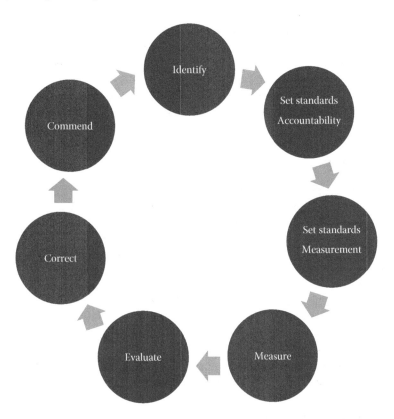

FIGURE 3.3   The ISSMEC management technique.

safety management process. As with all processes, existing targets and goals need to be achieved continually and new goals and objectives will arise from time to time.

RISK-BASED

The safety system must be a risk-based system. That means it must be aligned to the risks arising out of the workplace. Emphasis on certain system elements will be different according to the hazards associated with the work and the processes used. There is unfortunately no safety management system that will be ideal for all mines, industries and other workplaces, therefore they should be seen as a framework on which to build a risk-specific system for the industry. The main aim of the system is to reduce risks therefore the system must be aligned to those risks.

MANAGEMENT-LED

The key factor in safety and health management systems is management leadership. The safety system must be initiated and supported by senior management as well as line management. Only management has the authority and ability to create a safe and healthy workplace. This should be one of their prime concerns.

Safety systems that originate in, and which are maintained by the safety department, will have little effect on the organization. It is estimated that about 15% of a company's problems can be controlled by employees, but 85% can be controlled by management. Most safety problems are, therefore, management problems. If managers can manage the intricate and difficult concept of safety, then they will be able to manage other aspects of management easier, as managing safety enables them to be more effective.

AUDIT-DRIVEN

What gets measured usually gets done. Safety is an intangible concept and is traditionally measured after the fact—once a loss has occurred. The safety system must be an audit-driven system which calls for ongoing measurement against the standards and quantification of the results, before an accident.

A safety system converts safety intended actions into proactive activities and assigns responsibility and accountability for those actions, very similar to what a manager does with his or her subordinates. Each activity included in the safety system elements can be scored on a 1–5 scale to determine whether it has been achieved or not. At the end of the day, the entire system can be quantified by the score allocated. The safety system's effectiveness has been measured. The safety system elements and processes that scored less than full points are highlighted as areas that need improvement. Managers tend to pay more attention to processes that can be measured and quantified, and what gets measured gets done.

SAFETY MANAGEMENT AUDIT SYSTEMS

Formal safety management systems were initially seen as audit tools. In reality, an audit is difficult to conduct unless there are standards to audit against, and with no formal safety system in place, audits become baseline measurements rather than measurements of safety management efficiency, hence the terminology *Safety Audit Systems*.

SECTIONS, COMPONENTS, AND ELEMENTS

A safety management system comprises a number of ongoing processes and activities linked under a number of main groupings, sections, or headings. *Management Leadership* is an example of a section of a safety management system. One of the elements under this section would be, for example, management plant inspections. A sub-element would then dictate the minimum standards for the element and a further level would give the detail of the minimum standard. For example:

- Section: *Management leadership*
- Element: *Management inspections*
- Minimum standard: *Weekly plant inspections are to be done by all line managers and supervisors*
- Minimum standard detail: *The inspection checklist is to be completed in full during each inspection*

The Guidelines discussed in this publication normally give direction by listing the sections and elements of safety management systems but do not specify minimum standards or details, as these differ widely from plant to plant.

INTERNATIONALLY USED SAFETY MANAGEMENT SYSTEMS

There are a number of prominent international safety management systems currently in use throughout the world. Among others are: the National Occupational Safety Association (NOSA) 5-Star Safety Management System, the International Loss Control Institute (ILCI), International Safety Rating System (ISRS), the Occupational Safety and Health Administration (OSHA), Volunteer Protection Program (VPP), the International Labor Organization's (ILO) Guidelines on Occupational Safety and Health Management Systems, and many more. Many other programs are also available, but most address one or more elements of a safety system and not necessarily an entire safety management process.

THE NOSA 5-STAR SAFETY AND HEALTH MANAGEMENT SYSTEM

The 5-Star Safety and Health Management System was developed by the National Occupational Safety Association (NOSA) in South Africa in about 1968. The original score sheet was developed to generate interest in safety among a group of timber mills and sites. A checklist of 20–30 high-risk elements found at lumber sites was compiled, and a score (or weighting) was allocated to each element, so that by scoring on the checklist, they could compare upstream safety efforts on their various sites. The original idea was to have a competition between sites but not use the injury rate as the only measurement. The score allocated after the inspection and ranking would indicate to them their safer sites.

This system was so successful that NOSA started conducting safety surveys based on this system at mines, industries and commerce. After 50,000 safety surveys and inspections, more elements were added and the system was then promoted among industries and mines throughout South Africa. Around 1970, the system comprised 40 elements grouped under 5 main sections with a total score of 2000 points.

Upstream Safety Effort

This was one of the first safety systems to measure proactive, upstream safety efforts. It was used extensively to grade safety program efforts and recognize them on a national and international scale. A similar system was being developed in around 1978 in the USA, by Frank E. Bird of the International Loss Control Institute (ILCI), Atlanta, Georgia, in the form of the International Safety Rating System (ISRS).

In the 1980s, Frank Bird visited and consulted in South Africa and exchanged ideas concerning both the ISRS and NOSA Systems, which were way ahead of their time, as they were risk-based, management-led, and audit-driven systems. An exchange of expertise took place between ILCI and NOSA and the 40-element NOSA 5-Star Safety Management System was upgraded to 76 elements to be more in line with the ISRS.

Initially, NOSA awarded companies a rating, or grading of their safety systems, in the form of an A, B, or C grade based on the audit score. This proved very popular as organizations could now be recognized for their safety efforts based on management work being done to prevent loss, and not just on injury rates.

With the advent of the hotel star grading system around 1970, NOSA linked the score to a star grading based on the SEE audit formula (Safety Effort and Experience). A score was allocated for the safety effort (safety system effectiveness) and in conjunction with the injury rate (experience), a star grading from 1–5 was awarded. Challenging organizations to achieve a 5-Star grading for safety effort proved to be highly successful and eventually the NOSA 5-Star Safety System was exported to, and used in no fewer than 14 countries.

Management by Objectives

The NOSA 5-Star System was originally based on the concept of Management by Objectives, and was founded on solid management principles including the famous Mayo Hawthorne experiment. NOSA experienced the same results as Mayo did with their approach to safety by focusing on the motivational and recognition aspect of safety, rather than the punitive aspects. Awarding organizations a star grading based on the efficiency of their safety management system created competition, recognition, and spurred organizations on to improve their safety efforts. Many management principles, including those of Louis A. Allen and Peter Drucker are intertwined in the philosophy of the NOSA 5-Star System. The Loss Causation Model originally developed by W.H. Heinrich was modified in 1982 and used by NOSA as the basis of the 5-Star System philosophy.

The system is customized by the organization that uses it to adapt the principles to their specific hazards. The NOSA 5-Star standards cannot be reduced but can be increased, modified, and customized to suit the organization's specific risks (The NOSA 5-Star System, 1995).

The DNV GL International Safety Rating System (ISRS)

A widely used system, which sets standards of measurement, is the International Loss Control Institute's International Safety Rating System (ISRS), now owned and managed by DNV GL. This system has also set standards of conformance as well as standards of accountability for safety work. It incorporates physical inspection guidelines and precise instructions to accredited auditors on how to allocate scores. The ISRS eighth edition consists of 15 key processes, embedded in a continual improvement loop. Each process contains sub-processes and questions.

An ISRS assessment (audit) is a thorough evaluation of these questions and involves interviews with organizations where the questions are scored and commented. The scope of the assessment is entirely flexible and is determined by the size and complexity of the organization and the management's requirements. Detailed verification is conducted and organizations must be prepared to offer evidence to support their answers. The process' scores determine an overall level of performance between one and ten. The results provide a detailed measure of performance and gap analysis against the organization's desired level of performance. This becomes the planning basis for improvement during the next period. The ISRS eighth edition is structured with 15 processes embedded in a continuous improvement loop. These include:

---

1. Leadership	6. Project management	11. Contractor management and
2. Planning and	7. Training and competence	purchasing
administration	8. Communication and	12. Emergency preparedness
3. Risk evaluation	promotion	13. Learning from events
4. Human resources	9. Risk control	14. Risk monitoring
5. Compliance assurance	10. Asset management	15. Results and review

(DNV GL website, 2015)

These two pioneering auditing systems (NOSA 5-Star System and the ISRS) are both systematic and thorough in their approach and accurately quantify the work being done to control loss, while giving recognition for the safety effort. There are other similar systems in use, which operate in a similar way.

THE BRITISH SAFETY COUNCIL 5-STAR HEALTH AND SAFETY AUDIT SYSTEM (BSC)

According to the British Safety Council website:

> The audit is best suited to organizations that require a detailed and objective evaluation of their occupational health and safety management system(s) and associated arrangements. The audit measures performance against key safety management indicators, identifies areas of positive/negative practice and provides a structured approach for continuous improvement. The audit specification model includes two important safety management indicators which are continually assessed throughout the audit process:
> - Leadership (and commitment) at all levels of the organization
> - Continuous improvement
>
> There is also emphasis on the organization's approach to occupational health, employee well-being, safety culture, allocation of resources to health and safety and planning for change, all of which are considered as important best practice factors. (British Safety Council website, 2016a)

The British Safety Council describes the process as:

> The Five Star Audit process involves an in-depth examination of an organization's entire health and safety management system(s) and associated arrangements, focusing on the key aspects of their approach to managing occupational health and safety in the workplace and offers a structured path for continuous improvement towards best practice. The British Safety Council emphasizes that the five star audit specification model is reviewed and updated annually to ensure best practice health and safety management technique and trends are reflected within the content. (British Safety Council website, 2016a)

The BSC *Five Star Occupational Health and Safety Audit Specification* document, 2015, explains the elements and scoring methodology. The Five Star Audit process focuses on five sections and two Safety Management Indicators (SMI):

Section 1: Policy and Organization 750 Points
Section 2: Strategy and Planning 1125 Points
Section 3: Implementation and Operation 1500 Points
Section 4: Performance Measurement 1125 Points
Section 5: Evaluation and Review 500 Points
Total: 5000 Points

Safety Management Indicator (SMI) 1 Leadership Additional 0.5%–2% on overall Safety Management Indicator (SMI) 2 Continual Improvement.

The five sections of the audit are divided into 66 elements, which attract a maximum numerical value of 5000 points. Several of these elements are considered as "core" to the relevant section, and some of these core elements are also applicable within more than one section. Wherever an element of the audit is not applicable to the organization, it is withdrawn from the audit. The Maximum Accredited Audit Figure (MAAF) is the maximum total score available when non applicable questions have been removed and the Actual Accredited Audit Figure (AAAF) is the score achieved against such applicable questions.

Throughout the audit, two safety management indicators (sub-sections of leadership and continual improvement) are evaluated either as elements within their own right, in certain sections, or alternatively as scoring areas within other elements.

The cumulative scoring for these two sub-sections is then converted into a percentage figure, which can provide an additional 0.5%–2% to the overall audit grading. This aspect of the audit process is designed to encourage organizations to focus upon continually developing their safety management systems and culture through demonstration of commitment and robust leadership at all levels (British Safety Council, 2015b, p. 3).

SPECIALIZED SAFETY SYSTEMS

There are also a number of specialized safety systems which are unique to specific industries such as the System Safety methodology and the Process Safety System for the petroleum industry.

SYSTEM SAFETY

The system safety concept calls for a risk management strategy based on identification, analysis of hazards, and application of remedial controls using a systems-based approach. The systems-based approach to safety requires the application of scientific, technical, and managerial skills to hazard identification, hazard analysis, elimination, control, and management of hazards throughout the life cycle of a system.

Process Safety Management (PSM)

Process safety management is an analytical tool focused on preventing releases of any substance defined as a "highly hazardous chemical" by the Environmental Protection Agency (EPA) or the Occupational Safety and Health Administration (OSHA). Process safety management (PSM) refers to a set of interrelated approaches to manage hazards related to the process industry. It is intended to reduce the frequency and severity of undesirable events resulting from releases of chemicals and other energy sources. The standards are composed of organizational and operational procedures, design guidance, audit programs, and many other methods.

SAFETY MANAGEMENT SYSTEM GUIDELINES

Occupational Safety and Health Administration (OSHA), Volunteer Protection Program (VPP)

According to OSHA:

> The Voluntary Protection Program (VPP) promotes effective worksite based safety and health. In the VPP, management, labor, and OSHA establish cooperative relationships at workplaces that have implemented a comprehensive safety and health management system. Approval into VPP is OSHA's official recognition of the outstanding efforts of employers and employees who have achieved exemplary occupational safety and health.
>
> The Voluntary Protection Program (VPP) recognizes employers and workers in the private industry and federal agencies who have implemented effective safety and health management systems and maintain injury and illness rates below national Bureau of Labor Statistics averages for their respective industries. In VPP, management, labor, and OSHA work cooperatively and proactively to prevent fatalities, injuries, and illnesses through a system focused on: hazard prevention and control; worksite analysis; training; and management commitment and worker involvement.

To participate, employers must submit an application to OSHA and undergo a rigorous onsite evaluation by a team of safety and health professionals. Union support is required for applicants represented by a bargaining unit.

The main elements of the VPP program are as follows:

- Management Leadership and Employee Involvement
- Worksite Analysis
- Hazard Prevention and Control
- Safety and Health Training
- Self-Assessment

The self-assessment requirements for the VPP request the participants to provide narrative for answers concerning their safety management system around the four main sections and sub-sections of their safety management system as follows:

The participants are asked to explain in narrative form, the effectiveness of each of the four elements (and their sub-elements) of their safety and health management system. These are as follows:

1. Management Leadership and Employee Involvement
   a. Management Commitment to Safety and Health Protection and to VPP Participation
   b. Policy
   c. Goals, Objectives, and Planning
   d. Visible Top Management Leadership
   e. Responsibility and Authority
   f. Line Accountability
   g. Resources
   h. Employee Involvement
   i. Contract Employee Coverage
   j. Written Safety and Health Management System
2. Work Site Analysis
   a. Hazard Analysis of Routine Jobs, Tasks, and Processes
   b. Hazard Analysis of Significant Changes, New Processes, and Non-Routine Tasks, including pre-use analysis and new baselines
   c. Routine Self-Inspections
   d. Hazard Reporting System for Employees
   e. Industrial Hygiene Program
   f. Investigation of Accidents and Near Misses
   g. Trend/Pattern Analysis
3. Hazard Prevention and Control
   a. Certified Professional Resources
   b. Hazard Elimination and Control Methods, Engineering Controls, Administrative Controls, Work Practice Controls and Hazard Control Programs, Safety and Health Rules and Disciplinary System, Personal Protective Equipment
   c. Process Safety Management (if applicable)
   d. Occupational Health Care Program
   e. Preventive/Predictive Maintenance
   f. Tracking of Hazard Correction
   g. Emergency Preparedness
4. Safety and Health Training
   a. Managers
   b. Supervisors
   c. Employees
   d. Emergencies

In practice, VPP sets performance-based criteria for a managed safety and health system, invites sites to apply, and then assesses applicants against these criteria. OSHA's verification includes an application review and a rigorous onsite evaluation by a team of OSHA safety and health experts (OSHA website, 2016a).

Occupational Health and Safety Management Systems Specification (BS OHSAS 18001)

British Standards (BS) OHSAS 18001 is the Occupational Health and Safety Management Systems Specification Guideline. It comprises two parts, 18001 and 18002 and embraces a number of other publications. Key principles include the following:

- Leadership and management—clear commitment
- Setting objectives—continual improvement
- Planning—hazard identification, risk assessment and risk control
- Competence—training and awareness
- Consultation and communication—all stakeholders
- Structure and responsibility—clear lines and definitions
- SMS—audit and review to monitor effectiveness

BS OHSAS 18001 is a guideline for a safety management system and offers specifications for such a system. A health and safety management system is a formalized approach to health and safety management through use of a framework that aids the identification and control of safety and health risks. Through routine monitoring, an organization checks compliance against its own documented safety (and health) management system (SMS), as well as legislative and regulatory compliance. A well-designed and operated safety management system reduces accident potential and improves the overall management processes of an organization. Implementation of any structured safety system demonstrates an organization's commitment to protection of health and safety in the workplace.

BS OHSAS 18001 was specifically developed with requirements of the ISO 9001 Quality Management System and ISO 14001 Environmental Management System in mind, allowing for ease of integration of management systems. OHSAS 18001 was created via a concerted effort from a number of the world's leading national standards bodies, certification bodies, and specialist consultancies. A main driver for this was to try to remove confusion in the workplace from the proliferation of certifiable OH&S Specifications (British Standards Institute website, 2015a).

American National Standards Institute (ANSI) Z10–2012

The ANSI Z10 is a guideline standard for occupational health and safety management systems designed for companies and was the first standard of its kind published in the United States. Past experience showed that structured safety management systems, based on sound management practices, made an improvement in organizations' safety and health performance.

ANSI Z10 is intended to compliment organizations that have implemented other ISO standards such as ISO 9001 for quality and ISO 14001 for environmental control. It is an entirely voluntary standard based on recognized management standards. It is intended to offer organizations a tool for continual improvement of their safety management systems.

ANSI Z10–2012

Occupational Health and Safety Management Systems (ANSI/AIHA/ASSE Z10–2012) has been added as an updated version of the original, and according to the ANSI website:

> This standard lays out the blueprint for improving health and safety performance in your organization while increasing productivity, financial performance and quality. Sections include *Leadership and Employee Participation; Planning, Implementation and Operation; Evaluation and Corrective Action and Management Review.* The appendices address roles and responsibilities, policy statements, assessment and prioritization, audits, and more. (ANSI website, 2015a)

INTERNATIONAL LABOR ORGANIZATION (ILO)

According to their website:

> The International Labor Organization (ILO) is the only tripartite United Nations (UN) agency with government, employer and worker representatives. This tripartite structure makes the ILO a unique forum in which the governments and the social partners of the economy of its 186 member states can freely and openly debate and elaborate labor standards and policies.
>
> The unique tripartite structure of the ILO gives an equal voice to workers, employers and governments to ensure that the views of the social partners are closely reflected in labor standards and in shaping policies and programs. The main aims of the ILO are to promote rights at work, encourage decent employment opportunities, enhance social protection and strengthen dialogue on work-related issues. (International Labor Organization website, 2015b)

International Labor Organization ILO–OSH 2001

The document ILO–OSH 2001, entitled *Guidelines on Occupational Safety and Health Management Systems*, was drafted by the International Labor Organization (ILO) in 2001, and the second edition was released in 2009.

> The Guidelines call for coherent policies to protect workers from occupational hazards and risks while improving productivity. The Guidelines present practical approaches and tools for assisting organizations, competent national institutions, employers, workers and health management systems, with the aim of reducing work-related injuries, ill health, diseases, and deaths.
>
> The Guidelines may be applied on two levels—national and organizational. At the national level, they provide for the establishment of a national framework for Occupational Safety and Health (OSH) management systems, preferably supported by national laws and regulations. They also provide precise information on developing voluntary arrangements to strengthen compliance with regulations and standards, which, in turn, lead to continual improvement of OSH performance.
>
> At the organizational level, the Guidelines encourage the integration of OSH management system elements as an important component of overall policy and management arrangements. Organizations, employers, owners, managerial staff, workers and their representatives are motivated in applying appropriate OSH management principles and methods to improve OSH performance.

Employers and competent national institutions are accountable for and have a duty to organize measures designed to ensure occupational safety and health. The implementation of these ILO Guidelines is one useful approach to fulfilling this responsibility. They are not legally binding and are not intended to replace national laws, regulations or accepted standards. Their application does not require certification.

The Guidelines recommend five main sections or headings and some twenty one elements. The headings are; *Policy, Organizing, Planning and Implementation, Evaluation* and *Action for Improvement*. (ILO, 2001c, pp. 1–2)

International Organization for Standardization (ISO) 45001—Occupational Health and Safety Management System—Requirements

According to the document, ISO *Briefing Notes on ISO 45001*, ISO is developing an Occupational Health and Safety (OHS) Management System Standard Guideline (ISO 45001), which is intended to enable organizations to manage their OHS risks and improve their OHS performance. The implementation of an OHS Management System will be a strategic decision for an organization that can be used to support its sustainability initiatives, ensuring people are safer and healthier and increase profitability at the same time.

This draft standard, inspired by the well-known OHSAS 18001, is designed to help companies and organizations around the world to ensure the health and safety of the people who work for them.

ISO 45001 is an international standard that specifies requirements for an Occupational Health and Safety (OH&S) Management System, with guidance for its use, to enable an organization to proactively improve its OHS performance in preventing injury and ill-health. ISO 45001 is intended to be applicable to any organization regardless of its size, type and nature. All of its requirements are intended to be integrated into an organization's own management processes.

ISO 45001 enables an organization, through its OHS management system, to integrate other aspects of health and safety, such as worker wellness/well-being; however, it should be noted that an organization can be required by applicable legal requirements to also address such issues. (ISO website, 2015a, p. 2)

QUALITY, ENVIRONMENT, AND RISK MANAGEMENT STANDARDS

THE INTERNATIONAL ORGANIZATION FOR STANDARDIZATION (ISO)

ISO, the International Organization for Standardization, is an independent, non-governmental international organization with a membership of 162 national standards bodies. Through its members, it brings together experts to share knowledge and develop voluntary, consensus-based, market relevant International Standards that support innovation and provide solutions to global challenges. It is the world's largest developer of voluntary international standards and facilitates world trade by providing common standards between nations. (ISO website, 2016b)

ISO 9001:2015 QUALITY MANAGEMENT STANDARD

ISO 9000, Quality Management, was first published in 1987. It was based on the British Standards Institute BS 5750 series of standards from BSI that were proposed to ISO in 1979. An ISO 9001 quality management system assists organizations to continually monitor and manage quality throughout all of its operations. It is the world's most widely recognized quality management standard, and outlines methods and processes to achieve consistent performance and service.

According to the ISO booklet, *Quality Management Principles*, ISO 9001:2015 sets out the criteria for a quality management system and is the only standard in the family that can be certified. It can be used by any organization, large or small, regardless of the nature of its business. There are in excess of one million companies and organizations in over 170 countries certified to ISO 9001.

This standard is based on a number of quality management principles including a strong customer focus, the leadership and implication of top management, the process approach and ongoing improvement. Implementing ISO 9001:2015 helps ensure that customers get consistent, good quality products and services, which in turn brings many business benefits to both customers and the organization.

ISO 9001:2015, consists of 8 main sections which contain the following requirements:

- Section 1: Scope
- Section 2: Normative Reference
- Section 3: Terms and Definitions
- Section 4: Quality Management System
- Section 5: Management Responsibility
- Section 6: Resource Management
- Section 7: Product Realization
- Section 8: Measurement, Analysis, and Improvement

The standard specifies that the organization shall issue and maintain the following six documented procedures:

- Control of Documents
- Control of Records
- Internal Audits
- Control of Non-conforming Product/Service
- Corrective Action
- Preventive Action

In addition to these procedures, ISO 9001:2015 requires the organization to document any other procedures required for its effective operation and also requires the organization to issue and communicate a documented quality policy, a quality manual, and numerous records. The standard is based on eight quality management principles, which are defined in ISO 9000:2005. These are as follows:

Principle 1—Customer focus
Principle 2—Leadership
Principle 3—Involvement of people

Principle 4—Process approach
Principle 5—System approach to management
Principle 6—Continual improvement
Principle 7—Factual approach to decision making
Principle 8—Mutually beneficial supplier relationships (ISO, 2015c, pp. 1–7)

ISO 1400:2015 ENVIRONMENT STANDARD

According to the ISO (International Organization for Standardization) publication, *ISO 14001 Key Benefits*:

ISO 14001 is an internationally agreed standard that sets out the requirements for an environmental management system. It helps organizations improve their environmental performance through more efficient use of resources and reduction of waste, gaining a competitive advantage and the trust of stakeholders.

An environmental management system helps organizations identify, manage, monitor and control their environmental issues in a "holistic" manner. Other ISO standards that look at different types of management systems, such as ISO 9001 for quality management and ISO 45001 for occupational health and safety, all use a High-Level Structure. This means that ISO 14001 can be integrated easily into any existing ISO management system. ISO 14001 is suitable for organizations of all types and sizes, be they private, not-for-profit, or governmental. It requires that an organization considers all environmental issues relevant to its operations, such as air pollution, water and sewage issues, waste management, soil contamination, climate change mitigation and adaptation, and resource use and efficiency.

Like all ISO management system standards, ISO 14001 includes the need for continual improvement of an organization's systems and approach to environmental concerns. The standard has recently been revised, with key improvements such as the increased prominence of environmental management within the organization's strategic planning processes, greater input from leadership, and a stronger commitment to proactive initiatives that boost environmental performance. (ISO, 2016d, p. 2)

ISO 31000:2009 RISK MANAGEMENT STANDARD

The ISO website states the following:

ISO 31000: 2009, *Risk Management—Principles and Guidelines*, provides principles, framework and a process for managing risk. It can be used by any organization regardless of its size, activity or sector. Using ISO 31000 can help organizations increase the likelihood of achieving objectives, improve the identification of opportunities and threats and effectively allocate and use resources for risk treatment.

However, ISO 31000 cannot be used for certification purposes, but does provide guidance for internal or external audit programs. Organizations using it can compare their risk management practices with an internationally recognized benchmark, providing sound principles for effective management and corporate governance. (ISO website, 2009e, pp. 1–2)

HSE AND SHE OR EHS?

A safety management system follows a structured process for its implementation and maintenance. This is a similar process to quality control and environmental management processes. As proposed by the ISO, many organizations combine the functions of safety and health with that of environmental management, hence the terms SHE (Safety, Health and Environment) or HSE (Health, Safety and Environment) or EHS, (Environment, Health and Safety). Some organization's safety systems are termed HSE, SHE or EHS systems as are the coordinator's titles, for example, HSE Coordinator, SHE Department, etc.

Ideally, true safety professionals should focus on the safety system, qualified industrial hygienists on the industrial hygiene, and environmental specialists on the environmental protection system. While acknowledging that quality and environmental protection systems are management processes as well, this publication focuses only on the safety and health management system.

4 Risk-based Safety Management Systems

FAILURE TO ASSESS THE RISK

Failure to identify the hazards and to analyze and evaluate the risks and set up control measures, triggers off a chain of events that lead to accidental loss. Hazard Identification and Risk Assessment (HIRA) are the first steps in the prevention of accidental loss.

DEFINITION

A *risk* can be defined as *any probability, likelihood, or chance of loss*. It is the likelihood of an undesired event occurring at a certain time under certain circumstances. The two major types of risks are speculative risks, where there is the possibility of both gain and loss, and pure risks, which offer only the prospect of loss.

CONTROL

Failure to identify the risks brought about by the business as a result of having no safety management system, or having a weak system, leaves the elements and processes of a system that need to be controlled, unidentified. With no system or process to identify and reduce risks, accidental losses occur. This is an example of poor management control.

As a result of poor controls, root causes in the form of personal and job factors can arise, and can lead to high-risk work conditions and high-risk acts. These may eventually lead to a contact with a source of energy and a loss in the form of property damage or injury. This chain of events culminates in financial losses.

RISK MANAGEMENT

Risks within an organization cannot be properly managed until they have been assessed. Risk management combines the safety management functions of safety planning, organizing, leading, and controlling of the activities of a business, so as to minimize accidental losses and their adverse effects, produced by the risks arising from the operations of the organization.

Physical risk management consists of identifying the hazards, assessing the risks, evaluating the risks (Risk assessment), and introducing the necessary controls to reduce the probability of these risks manifesting in loss.

RISK ASSESSMENT

Risk assessment is a method that is predictive and can identify potential for loss. With this knowledge, an organization is able to set up the necessary management controls in the form of safety management system processes, to prevent these risks resulting in losses such as injuries, property damage, business interruptions, and environmental harm. Many safety systems focus on the consequence of loss, and not the control which renders them ineffective. Effective risk assessment is proactive, predictive safety in the finest form. In risk assessment, the keywords are, "It's not what happened, but what *could* have happened."

DEFINITION

Risk Assessment can be defined as: *the evaluation and quantification of the likelihood of undesired events and the likelihood of injury and damage that could be caused by the risks.* It involves an estimation of the consequences of undesired events.

One of the biggest benefits of risk assessment is that it will indicate where the greatest gains can be made with the least amount of effort, and which activities should be given priority during the implementation of the safety management system. The organization can now draw up a prioritization system based on sound assessments of its risks.

COMPONENTS OF RISK ASSESSMENT

The three major components of risk assessment are as follows:

1. Hazard Identification
2. Risk Analysis
3. Risk Evaluation

Once the three phases of risk assessment are completed, risk controls can then be implemented. Risk control can only be instituted once all hazards have been identified, and all risks quantified and evaluated. The risk assessment approach is a systematic approach to workplace health and safety, and focuses on the bigger picture. On the other hand, traditional approaches to safety improvements were aimed at eliminating the high-risk acts and unsafe work conditions (symptoms) only, while missing the real causation factors, namely, the accident root causes (causes).

HAZARD IDENTIFICATION

The range of activities undertaken by an organization will create hazards, which will vary in nature and significance. The range, nature, distribution, and significance of the hazards is called the *hazard burden* and will determine the risks which need to be controlled.

HAZARD BURDEN

Ideally, the hazard should be eliminated altogether, either by the introduction of inherently safer processes or by no longer carrying out a particular activity. This is not always practical. If the hazard burden is reduced and if other variables remain constant, including consistent operation of the safety management system, this will result in lower overall risk and a consequent reduction of undesired events and consequent injuries, ill health, or damage.

DEFINITION

A *hazard* can be defined as *a situation which has potential for injury, damage to property, harm to the environment, or a combination of all three.* A hazard is a source of potential harm and high-risk acts, and high-risk work conditions are examples of hazards.

Some of the hazard classifications are as follows:

- Biological—bacteria, viruses, insects, plants, birds, animals, and humans, etc.
- Chemical—depends on the physical, chemical, and toxic properties of the chemical
- Ergonomic—repetitive movements, improper workstation set up, etc.
- Physical—radiation, magnetic fields, pressure extremes (high pressure or vacuum), noise, etc.
- Psychosocial—stress, violence, etc.
- Safety—slipping and tripping hazards, poor machine guarding, equipment malfunctions or breakdowns

HAZARD PRIORITIZATION

Over the years, a lot has been written concerning various hazard classification methods, terminology, and order of priority. The main reason for hazard classification is to give priority to the hazards that have the highest potential for loss to people, equipment, machinery, or the environment. All hazards identified should ideally receive priority and should be rectified immediately wherever possible. Unfortunately, due to limited manpower and limited resources, not all hazards can be rectified immediately, and consequently a classification system is necessary to give priority to the most critical hazards.

A simple hazard classification system is the A, B, and C classification:

A. Likely to cause death, permanent disability, extensive property damage, or even catastrophic results
B. Likely to cause serious injury but less serious than an A class hazard, substantial property loss, or damage to the environment
C. Likely to cause minor injury, relative property damage, and minor disruption

During classification, workplace hazards which include practices or conditions that release uncontrolled energy (contact and exchange of energy), should be considered. Examples are as follows:

- An object that could fall from a height (potential or gravitational energy)
- A runaway chemical reaction (chemical energy)
- The release of compressed gas or steam (pressure; high temperature)
- Entanglement of hair or clothing in rotating equipment (kinetic energy)
- Contact with electrodes of a battery or capacitor (electrical energy)

When considering the hazard classification, the *consequence* of the outcome of the hazard should be considered as well as the *probability* of an undesired event occurring. Taking these two factors into consideration gives one an immediate risk analysis of the particular hazard and assists in prioritizing its rectification.

Hazard Profiling

Hazard profiling is a process of describing the hazard in its local context, which includes a general description of the hazard, a local historical background of the hazard, local vulnerability, possible consequences, and estimated likelihood, and is very similar to risk assessment.

HAZARD IDENTIFICATION METHODS

The first step of a risk assessment is the identification of all hazards within an organization. There are numerous hazard identification methods and techniques. The two main techniques are the *fundamental* and the *comparative* methods. *Fundamental* techniques include the following:

- Hazard and Operability studies (HAZOP)
- Failure Mode and Effect Analysis (FMEA)
- Failure Mode, Effect, and Critical Analysis (FMECA)
- So What if it Happens (SWIFT)
- Event and Fault Tree Analysis
- Past Accidents and Near Miss Incidents
- Lessons Learned
- Single Root Cause Analysis
- Critical Task Identification
- Safety System Audits
- Brainstorming
- Delphi Technique
- Human Reliability Analysis

Comparative techniques use checklists based on industry standards or existing codes of practice. They could involve comparing the plant in question with similar plants.

HAZOP (Hazard and Operability Study)

A hazard and operability study (HAZOP), is a structured and systematic examination of a planned or existing process or operation in an organization, in order to identify and evaluate items that may represent risks to personnel or equipment. A HAZOP is carried out by a suitably experienced multi-disciplinary team during a set of meetings where aspects of a plant are subject to guidewords applied to relevant physical properties, such as flow, temperature, and pressure. A potential cause of a deviation is then sought, and a consequence is defined. If the consequence is undesirable and losses could occur, then the hazard has to be addressed by removal, mitigation, or control. Mitigating action would be actioned with responsibility delegated.

The HAZOP technique was initially developed to analyze chemical process systems and mining operation processes, but has later been extended to other types of systems and complex operations such as nuclear power plant operations.

FMEA (Failure Mode and Effect Analysis)

FMEA is another method of hazard identification (Figure 4.1). This method is used for identifying possible failures in the system and resulting consequences. FMEA asks the question, "What system could fail and what would be the consequence?"

An example of a FMEA method of hazard identification is given in the figure below, which shows extracts of an analysis done in a power generating unit. The FMEA exercise identified the main systems, which could fail within the department, and the consequences as a result of main system failures (Figure 4.1). FMEA asks, "What can fail and what will the effect be?"

FMECA (Failure Mode, Effect, and Criticality Analysis)

The Failure Mode, Effect, and Criticality Analysis (FMECA), is a hazard identification method that goes into more depth than the FMEA. The FMECA method examines each component of a system for criticality and identifies the effect on the entire system upon failure of components. This helps focus on critical components within a unit.

SWIFT (So What If It Happens?)

A SWIFT analysis is a structured system for prompting a selected team to identify risks and possible consequence analysis. It is normally used within a facilitated

ITEM UNDER REVIEW	WHAT COULD FAIL?	WHAT EFFECT?
Generator	The generator bearings overheat	Slow speed, bearing seizure, damage
Main transformer	The windings could fail	No power transmission, circuit damage

FIGURE 4.1 Failure Mode and Effect Analysis (FMEA).

workshop. It can be linked to a risk analysis and evaluation technique, and asks for answers on identified scenarios based on the question, "So what if it happens?"

EVENT TREE AND FAULT TREE ANALYSIS

The Event Tree Analysis (ETA) is a predictive method of determining the cause and effect of events. The ETA starts with the event and deduces by means of Boolean logic, what factors could contribute to the event. Fault Tree Analysis (FTA) is deductive as it deduces the events and sub-events that lead to the main event using the same method as Event Tree Analysis.

PAST ACCIDENTS AND NEAR MISS INCIDENTS

A useful method of predicting future hazards, is to review past injury and property damage causing accidents as well as near miss events. By studying past loss-producing events, a pattern can be derived that would indicate certain recurring and inherent hazards within the business.

Near miss incidents, or events, which under slightly different circumstances could have resulted in a loss, are perhaps the best indicators of the presence of hazards arising from the risks of the business. They can often highlight hazards that can then be controlled before the loss occurs. Formal and informal incident recall sessions are imperative if hazard identification is to be done thoroughly. Incident recall is a method whereby employees recall past near miss incidents that under slightly different circumstances could have resulted in accidental loss. The losses could cause injury to people, property damage, or interruption to the work process. Near miss incidents are vital in hazard identification. Near miss incident reporting systems have often failed in the past, yet can be very successful if they are an integral component of the safety management system.

LESSONS LEARNED

Learning from accidents (lessons learned) is another method of identifying hazards, which is similar to incident recall, and is facilitated by asking the following questions:

- What hazardous events have occurred in the past, within the organization or in the industry as a whole, in facilities producing the same or similar product using the same or similar process?
- Can these events or similar events occur in the process under consideration?
- What lessons have been learnt?

SINGLE ROOT CAUSE ANALYSIS (SINGLE LOSS ANALYSIS)

A single loss that has occurred is analyzed in order to understand contributory causes and how the system or process can be improved to avoid such future losses.

The analysis considers what controls were in place at the time the loss occurred and how controls might be improved. This is a part of the "lessons learned procedure."

CRITICAL TASK IDENTIFICATION

A systematic method of listing work tasks and applying a critical task hazard analysis will highlight hazards or hazardous steps within certain tasks. This will enable written safe work procedures to be written for these critical tasks. These will help guide the employee to avoid these risks when carrying out the critical tasks. Past injury and loss experience, as well as probabilities are examined. Near miss incidents are also considered in the critical task evaluation.

SAFETY SYSTEM AUDITS

Structured safety audits of the existing safety management system also indicate what elements and items are not in control and which pose a hazard. These audits point out a company's strengths and weaknesses and also indicate which areas may possess more hazards.

BRAINSTORMING

Brainstorming is a means of collecting a broad set of ideas, evaluating and ranking them by a team and selecting solutions or conclusions. Brainstorming may be stimulated by prompts or by one-on-one and one-on-many interview techniques.

DELPHI TECHNIQUE

This is a means of combining expert opinions that may support the source and influence identification, probability, and consequence estimation and risk evaluation. It is a collaborative technique for building consensus among experts and involves independent analysis and voting by subject experts.

HUMAN RELIABILITY ANALYSIS (HRA)

HRA deals with the impact of humans on system performance and can be used to evaluate human errors and how they influence a system.

SAFETY INSPECTIONS

A thorough safety inspection guided by a hazard control checklist is one of the basic, best, and widely used methods of identifying physical hazards as well as high-risk practices. The hazards, once identified, should be ranked as to their classification using a simple system such as the A, B, and C rating method.

All work areas should be inspected for hazard recognition. All machinery, tools, and equipment used should be included in the inspection, and environmental conditions should be sampled and tested. These would include the level of illumination, ventilation, as well as noise levels, etc.

DEFINITION

A safety inspection is a monitoring function to endeavor to locate, identify, and eradicate unsafe conditions (hazards) and high-risk acts, which have the capacity to lead to accidental loss in the work area.

An inspection involves a tour around the physical work environment with the specific objective of ensuring the safety of the people, products, equipment, and machinery in that area. Inspections should be an integrated part of each manager's normal duty and the main purpose of a safety inspection is fact-finding, not fault-finding.

PURPOSE OF INSPECTIONS

The main purpose of inspections is to identify hazards that have the potential to cause loss. All risks or chances of loss can be identified during an inspection and sub-standard work conditions can be identified and noted for corrective action.

The inspection of the physical workplace, not only includes the work area itself, but also the process being carried out, the movement of material, raw product, and finished goods as well as the actions, working conditions, and general safety of the employees. Short cuts, high-risk behavior and processes can also be detected during an inspection, and once these facts are identified, positive remedial measures can be taken to rectify them as part of the proactive safety process.

Inspections help detect hazards and are one of the most important precontact control mechanisms of a safety management system. *"What gets measured, gets done."* Consequently, a safety inspection is the ideal method of measuring the physical condition standards and work procedure standards in operation in the plant against the safety system standards. The effective way of measuring compliance to standards is by conducting a safety inspection, often referred to as *management by walkabout*, and is part of visible felt management leadership.

Inspections can be conducted over and over again with no impact, or benefit, if the deviation from standards found during the inspection is not rectified. Regular inspections act as follow-up tours for actions that were detected during the previous inspection and which should have been rectified.

WHERE TO INSPECT

Often only the prime work areas are included in safety inspections. Although these areas may pose more of a threat to the workers in that particular area, all work areas must be inspected regularly. The frequency of the inspections will be determined by the hazardous nature of the work environment. Also to be considered is the process, and the potential for major loss in the form of injury, fire explosion, or other risks.

A thorough safety inspection should cover the entire work area, including office buildings and everything within the boundary fences. Detailed inspections should be carried out in workshops, canteens, change rooms, storage areas, machinery installations, inside and on top of cupboards, cabinets, and inside lockers. Cabs and loading bays of vehicles, as well as cabs of overhead cranes, fork-lift trucks, and other motorized equipment should also be inspected.

TYPES OF INSPECTIONS

There are numerous classifications of safety inspections. The classifications may also vary from place to place, industry to industry, and country to country. A few types of safety inspections that can be carried out are listed here and are intended as a guide rather than a comprehensive list.

Risk Assessment Precursor

A risk assessment cannot be carried out successfully without first of all conducting an on-site inspection of the physical conditions, the raw products/materials used, the processes, and machinery as well as the transportation routes, be it motorized or mechanical. Only once the entire process and exposures have been examined by an inspection, can a thorough risk assessment of the most hazardous situations be compiled. A risk assessment without some form of physical hazard identification inspection is virtually impossible.

Regulatory Compliance

A legal compliance audit involves an inspection to confirm compliance to regulations and laws governing the physical work conditions, as well as the control systems, and conformance to legal reporting norms and other aspects. This inspection will inspect the work areas and installations and may include an examination of documents, inspection of safety records, report forms, records of training given, as well as various appointments under different regulations.

Third Party Inspections

Another type of safety inspection is an inspection carried out by an external risk or safety consultant. The inspection may be as a result of certain problems, on invitation by the organization, or an annual pre-requisite prior to being assessed for safety performance achievements, or insurance purposes. An outside consultant may have been called in as a result of a total collapse of the safety system. The third party inspection may be a legal inspection by safety authorities.

Informal Safety Walkabout

An informal inspection should take place on a regular basis, preferably daily. These are sometimes referred to as management tours, but correctly termed, they are safety inspections. This would be a walkabout by the manager, supervisor, or foreman of that work area, and should cover all the areas under their jurisdiction. A lack of constant and regular informal safety inspections could lead to frequent occurrence of accidents as a result of high-risk acts and conditions.

An informal inspection can immediately identify these hazards and unsafe work practices, which can then be immediately rectified on a daily basis. This gives a tighter control over the causation factors of near miss incidents and accidents, making this an important inspection.

Planned Inspections

Planned inspections are those inspections that are planned on a regular basis. They normally follow a predetermined route and are carried out by a team comprising the Safety Coordinator, the Supervisor of the area, and possibly the Safety and Health Representative of that area.

These inspections are normally scheduled, a certain date is allocated (weekly, monthly, or three-monthly) and the areas nominated for inspection are determined long before the inspection. Planned inspections are an ideal way of covering the entire work area, as they are a systematic approach to covering every square foot of the factory, plant or mine, to identify hazards.

Safety Department Inspections

Being thoroughly familiar with the work environment and also being specialists in their field, the regular inspections carried out by the safety department personnel are of vital importance. It is also a systematic approach to ensuring that all hazards, deviations from laid-down standards, and sub-standard practices are identified and rectified on an ongoing basis.

The safety department staff should take the opportunity of having the local manager/supervisor accompany them on their inspections and should never inspect on their own. This could be the waste of an ideal opportunity to coach others in the safety inspection technique so that it becomes a learning experience.

During these inspections, the safety department personnel will also build up a lot of credibility. This is inclined to change the perception from a fault-finding inspection to a fact-finding inspection by the safety department and the local manager or supervisor.

Safety and Health Representatives' Inspections

Safety and Health Representatives are appointed because of their knowledge and familiarity with the work environment and work process. No one is better equipped to carry out a safety inspection of a particular area than the person who works in the area, knows the process and has many years' experience in their work environment. Safety and Health Representatives also prove invaluable, as their inspection techniques are very thorough and no quarter is overlooked during their inspections.

Although they may not have the authority to take remedial action on certain hazards noted, the Safety and Health Representatives' reports must receive prompt attention by management to prevent the same hazards from being reported over and over again. Once it appears that no actions are forthcoming from the Safety Representatives' inspection reports, they could become disillusioned, lose heart, and the benefits of their inspection system could be lost.

Involvement of, and inspections of areas by Safety and Health Representatives, is based on the principle of involvement in safety, which subsequently motivates the people that are involved. Constant reinforcement, support and commendations should be given to the Safety and Health Representatives concerning their inspections, and an ongoing system of encouragement should be in place.

Safety Surveys

A safety survey is a thorough safety inspection of the work environment and also includes an inspection of control systems in place. Normally, a safety survey is the first step in introducing a long-range safety and health management system, and would highlight the most prominent deficiencies in the physical conditions and control systems. The safety report sent out after the safety survey acts as an action list for management to initiate an ongoing safety management system.

Safety Audit Inspections

A safety audit inspection would be a thorough examination of the physical conditions, the equipment, machinery, and systems in place, measured against an established and quantifiable standard of measurement. The safety audit would verify, *what has been done, who did it, and the frequency with which it was done.*

Safety Review Inspections

Depending on the safety management system in operation at a plant, a review or evaluation of the entire system will include an inspection of the physical areas, the control systems, critical items of machinery and apparatus, and include control documents appertaining to control processes.

Specific Equipment Inspections

Specific equipment inspections are routine inspections of items of equipment which are used regularly and which are essential to the health, safety, and wellbeing of people. These could include critical parts and components. Special equipment inspections could include either daily, weekly, monthly, quarterly, or even annual inspections of such equipment, determined by the frequency of use and consequences of equipment failure. They could include firefighting equipment, electrical equipment, fixed electrical installations, new equipment, ladders, pressure vessels, boilers, lifting gear, etc.

OTHER TYPES OF INSPECTIONS

Safety inspections can vary greatly from workplace to workplace and could include inspections such as:

- Ergonomic inspections
- Accident/near miss incident investigation inspections
- Legal inspections by the local legal representative
- Occupational hygiene inspections
- Housekeeping inspections
- Safety competition inspections
- Inspections of a contractor's site
- Self-audit inspections

- Annual audit inspections
- Safety signage inspections
- Critical PPE inspections
- Risk management inspections by insurers, etc.

Depending on the nature of the industry, the list could be extended. The intention is to give an indication of the type of safety inspections that could be carried out periodically.

INSPECTION CHECKLISTS

An inspection checklist is a thought starter, a reminder, and a prompt for the inspector as to what he or she must look for during the inspection. During safety inspections, one can get distracted by the activity, noise, and flow created by the production lines. Some form of guideline is required to help concentrate the attention on the safety matters at hand.

The checklist also provides for a quick summary of the measurement. Deviations can be noted opposite the item and lengthy reports need not be made while on the inspection tour. The checklist also acts as a record of what has been inspected, it records the name of the inspector, the date of the inspection, and also the area being inspected. The safety inspection checklist structures the inspection and makes it more meaningful and thorough.

TRAINING OF INSPECTORS

Safety inspecting is a specialized task and, if possible, training should be given in correct inspection techniques. Normally, supervisors and production orientated employees, are mostly interested in the work process and the technicalities of the work area, and are not fine-tuned to inspect the safety aspects applicable in their particular areas.

An outside trained safety inspector who has attended some form of inspection technique training sessions will be far more orientated toward looking for, and highlighting deviations from accepted standards.

DANGER TAGS

It may not be possible to rectify a hazard immediately during the inspection, and an alternate method of warning others of the hazard would be acceptable in most instances. Numerous organizations use a danger tag or *unsafe to use* tag which is tied onto the piece of equipment or secured in the vicinity of the hazard in a visible position, warning people not to operate or work the equipment. These *do not use* tags are only a temporary measure and should not be viewed as having rectified the hazard.

A list of the tags used during the inspection should be made, and the report and follow-up of the inspection would indicate what attention should be given to which hazard. Bearing in mind that the accident root cause "failure to warn" is responsible for a number of injuries as a result of accidents. This tag provides a warning to employees that the equipment is unsafe and should not be used.

RISK ANALYSIS

Once hazards have been prioritized by a hazard-ranking exercise, which follows the hazard identification process, the next step in risk assessment is the analysis of the risks.

DEFINITION

Risk analysis is: *the calculation and quantification of the probability, the severity, and frequency of an undesired event occurring as a result of a hazard.* It is a systematic measurement of the degree of danger in an operation and is the product of the probability and frequency of the undesired event and the resultant severity. Many methods of risk analysis only consider the potential severity and probability of occurrence. The third dimension of the analysis, the frequency of occurrence, has been added in Figure 4.2 for clarification.

PURPOSE

The purpose of the risk analysis is to reduce the uncertainty of a potential accident situation and to provide a framework to look at all eventualities. A risk analysis is a risk quantification method that looks, not at what happened in the past, but what could happen in the future. It is a method of identifying accidental events that have not yet occurred.

Risk Score = Probability × Severity × Frequency						
PROBABILITY ⟶	Damage 1	Fire 2	Explosion 3	Injury 4	Severe injury 5	Fatality 6
X	X	X	X	X	X	X
SEVERITY ⟶	Minor Injury/Damage 1	Serious Injury/ Damage 2	Permanent Disability/ Disruptive Damage 3	Multiple Disability/ Major Damage 4	Fatality/ Catastrophic Damage 5	Multiple Fatalities/ Permanent Loss of Structure 6
X	X	X	X	X	X	X
FREQUENCY ⟶	Anytime 6	Hourly 5	Daily 4	Weekly 3	Monthly 2	Yearly 1
SCORE (P × S × F)	Example: If Risk B has a probability of 5, a severity of 4, and frequency of 4, that gives a score of 80					

FIGURE 4.2 Analyzing risks by weighing probability with severity and frequency.

RISK SCORE

The risk score is the product of the probability of the event's happening, the severity of the consequences, and the frequency of exposure to the event. The probability asks the question, "What are the chances of the event happening?" The severity asks the question, "If it happens, how bad will it be?" and the frequency asks, "If it happens, how often can it occur, and how many people are exposed?" Numbers are allocated to the various degrees of probability, severity, and frequency and the product gives a risk score.

In Figure 4.2, the probability, severity, and frequency are ranked on a scale of (1) to (6). Frequency, or exposure, has a ranking of (6) to (1). The highest probability with the worst severity and the most frequent exposure would be (6) multiplied by (6) multiplied by (6), which would be the highest risk score on the table of 216. The scores for the six different risks assessed are calculated by multiplying the applicable probability number with that of the severity and frequency. The scores help prioritize remedial actions. Example: Risk B has a probability of (5), a severity of (4), and frequency of (4) giving it a score of ($5 \times 4 \times 4 = 80$).

RISK MATRIX

A simple risk analysis of the hazard can be done using the risk matrix. For discussion purposes, a simplified risk matrix is used here. A risk matrix is a block diagram with two axes, the loss potential *severity* and *probability* of occurrence, both ranked low, medium, medium-high, high, and extreme. Figure 4.3 is an example of a simplified risk matrix that can be used for risk analysis.

Each level has been numbered and the axis is multiplied to give a score of the particular ranking. If the probability of occurrence was extreme (5) and the potential severity medium-high (3), a risk score of (15) would be obtained. All hazards should

Probability of occurrence	Low	Medium	Medium-high	High	Extreme
Extreme	5	10	15	20	25
High	4	8	12	16	20
Medium-high	3	6	9	12	15
Medium	2	4	6	8	10
Low	1	2	3	4	5

Potential severity

FIGURE 4.3   A simple risk matrix.

be risk assessed as some have more potential for loss than others and this can only be determined by the risk analysis exercise.

Each hazard is then ranked as to its potential severity and probability of occurrence. Hazards ranked in the (20–25) zone should receive immediate attention. The next priority are those that rank (16). Those in the (9–15) zone should be rectified as soon as possible. Hazards in the range of (4–10) and which do not pose an immediate threat to life or limb should be next on the priority list, and those ranking (1–5) should receive attention once the higher risks have been rectified.

RISK FREE?

For any organization to be absolutely risk free would be improbable as well as impracticable. In risk reduction, the objective should be to drive the risk As Low As is Reasonably Practicable (ALARP). The ALARP zone is where the risks have been reduced to where they can be tolerated and the risk level is acceptable. Certain risks have to be tolerated as long as they are kept in the ALARP zone, and the processes of the safety management system ensure, with ongoing checks and balances, that the risks remain as low as possible. If the risks are kept in the ALARP zone, it is regarded as normal business practice. Risks that extend above the ALARP zone could prove to be detrimental to the business as well as the employees, the public and others.

RISK PROFILE

Definition

A *risk profile* is: *a representation at a given point in time of an organization's overall exposure to some specific risks.* A desired risk profile is a target risk profile which the organization wishes to adhere to over time, based on its risk reduction objective, corporate governance approach, and overall strategic objectives. Management is thus tasked with continually adjusting the current risk profile through risk mitigation and risk transfer mechanisms, so as to align the organization with the target risk profile.

RISK EVALUATION

Risk evaluation could be defined as: *the quantification of the risks coupled with an evaluation of the cost of risk reduction and the resultant benefit derived from reducing the risk.* The main objective of risk evaluation is to ensure that the cost of risk reduction justifies the degree of risk reduction. Its main aim is to enable management to make decisions on risk reduction priorities based on a cost benefit analysis.

RISK CONTROL

Risks cannot be managed until they have been assessed. Once the risks have been assessed and prioritized through the risk evaluation process, risk controls are now implemented in the form of safety management system processes.

The objective of risk control is to minimize those risks that have been assessed and wherever possible to terminate or transfer them. Exposure to the risk should be minimized and then contingency plans should be drawn up to cope with the consequences should the event occur. The goals of risk control are to reduce the probability, severity, and frequency of undesired events occurring, to a level as low as practicable.

RISK CONTROL METHODS

There are basically four ways to handle risks:

1. Terminate the risk
2. Tolerate the risk
3. Transfer the risk
4. Treat the risk

Terminate

The ideal method of risk mitigation is to terminate the risk entirely. This would mean stopping a process, changing the business, or disposing of a substance used in the process, so that the risk is entirely terminated. In the hierarchy of control, this is referred to as Elimination, and is the preferred solution. This involves eliminating the risk entirely or designing the risk out of the process entirely. This is the most effective method of creating a safer work environment (Figure 4.4).

Tolerate

If risks are in the ALARP region, it is acceptable in business practice to tolerate these risks, as there is risk in all aspects of business. The tolerating of a risk means that the benefits derived from the risk outweigh the consequences of the risk. Tolerating the risk will involve the last three controls in the hierarchy, that is, Engineering Controls,

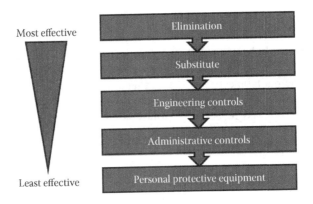

FIGURE 4.4   The safety hierarchy of hazard control.

Administrative Controls, and the providing of Personal Protective Equipment and barriers.

Transfer

Transferring the risk is not always an ideal solution to control risk and normally involves insuring the risk or placing it in somebody else's hands. This would be shifting the risk from one person or organization to another.

Treat

Treating the risk means setting up safety and health management controls to reduce the risk and therefore reduce the probability of an undesired event occurring. This would call, therefore, for the implementation and maintenance of a structured safety and health management system. Treating the risk would mean all the layers of the hierarchy of hazard control would be implemented and maintained by the safety management system.

Treating the risk within an organization involves the safety management principle of safety controlling. Safety controlling is the management function of identifying what must be done for safety, inspecting to verify completion of work, evaluating, and following up with safety action.

RISK REGISTER

A Risk Register is a record of details of all the risks identified within the company, their grading in terms of probability of occurring and severity of impact on the company, initial plans for mitigating each high level risk, the costs and responsibilities of the prescribed mitigation strategies, and subsequent results. It usually includes the following:

- A unique identifier number for each risk
- A description of each risk and how it could affect the organization
- A risk assessment of probability and severity (low, medium, medium-high, high, extreme)
- A risk ranking of each risk
- Who is responsible for managing the risk
- Proposed mitigation actions (preventative and contingency)
- Estimated costings for each mitigation strategy

The risk register should be maintained and updated by each department and will change regularly as existing risks are reviewed in the light of the effectiveness of the mitigation strategy, and new risks are identified.

PLAN, DO, CHECK, ACT (PDCA)

Risk treatment by implementing a safety management system can be based on the methodology known as *Plan, Do, Check, Act* (PDCA), which is a system of processes and is recommended by the Guidelines such as OHSAS 18001 and ISO 45001.

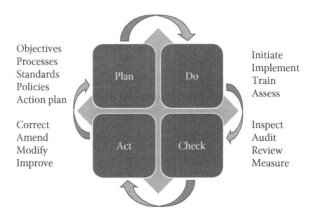

FIGURE 4.5   Plan, Do, Check, Act (PDCA) process.

A summary of the PDCA process is as follows:

Plan—establish the objectives, processes, systems, checks and balances in the form of safety system elements necessary to deliver results in accordance with the organization's safety and health policy, safety performance indicators and action plan.
Do—implement the safety management system and its processes, checks, balances, and elements according to the action plan.
Check—monitor and measure processes against the legal requirements, safety system standards, and action plan objectives using the management control function, which includes inspection and audits. The results must be reported and quantified.
Act—take corrective actions to continually improve the safety management system. [British Standards Institute (BSI) website, 2016b]

Sustainability is achieved by maintaining ongoing processes and improvements to ensure that a process of continual improvement to health and safety is in place (Figure 4.5).

THE SAFETY MANAGEMENT CONTROL FUNCTION

Once the risks have been assessed, they can be treated by applying safety management *control*, which integrates the PDCA process into the safety management function. The acronym IISSMECC is used to explain the safety management control function, where:

I = Identify and assess the risks. (Plan)
I = Identify the actions needed to reduce the risks. (Plan)
S = Set standards of accountability. (Do)
S = Set standards of measurement. (Do)
M = Measure against those standards. (Check)

E = Evaluate conformances and non-conformances. (Check)
C = Corrective action to be taken. (Act) (Ongoing improvement)
C = Commendation for work well done. (Act)

SUMMARY

Failure to assess, analyze, evaluate, and control risks is the key factor that triggers off the chain of events referred to as an accident that results in undesired losses such as injuries, fires, property damage, and business interruption. Risk assessment, risk management, and risk control is the foundation of the safety management system and will lead to predicting where loss occurring events may happen so that they can be prevented.

5 Management-led Safety Management Systems

INTRODUCTION

If the risks arising out of a business have not been identified and assessed, they cannot be managed or controlled. This creates a lack of, or poor management control in the form of a weak safety management system (SMS), no safety standards, or non-conformance to existing standards. This is depicted by the second domino in the chain of events leading to a loss.

PRINCIPLES OF SAFETY MANAGEMENT

The following principles will indicate the importance of management's role in safety management. The safety management system must have the full support of, and must be driven by management at all levels who must participate in the safety activities and as leaders must set the example.

MANAGEMENT LEADERSHIP

The safety and health of employees at a workplace is the ultimate responsibility of all levels of management within the organization. Even though it is generally accepted that all share a role in safety, the ultimate accountability lies heavier with all echelons of leadership. With this in mind, a safety management system can only be successful if initiated and led by top management and supported by line management.

Management Leadership in Occupational Safety and Health—A Practical Guide, published by the European Agency for Safety and Health at Work (EU-OSHA), lists three main leadership principles for the enhancement of occupational safety and health.

They are as follows:

- Effective and strong leadership
- Involving workers and their constructive engagement
- Ongoing assessment and review

EFFECTIVE AND STRONG LEADERSHIP

Leadership is a condition for success. A preventive approach is only likely to be fruitful if it is supported by the management. A strong and visible leadership and engaged managers at all levels can provide direction and input to this preventive approach.

This makes it clear to everyone that safety and health are strategic questions within the company. In practice, this means the following:

- Management commits itself to occupational safety and health as a core value of the organization and communicates this to the employees.
- Managers have an accurate picture of the risk profile of the organization.
- Management leads by example and demonstrates leadership integrity, for example, by following all occupational safety and health rules at all times.
- The roles and responsibilities of different actors in preventing and managing health and safety risks at the workplace are clearly defined and planned and monitored proactively.

ONGOING ASSESSMENT AND REVIEW

Monitoring and reporting are vital tools for enhancing workplace safety and health. Management systems which provide, for example, the company board members with specific and routine reports on the performance of health and safety policies can be useful in raising issues, highlighting problems, and ultimately enhancing workplace safety and health. Elements of a good assessment and review system include the following:

- Procedures for reporting major health and safety failures to the company board members and owners as soon as is practical
- Systems for capturing and reporting accurate and timely incident data such as accident and sickness rates
- Arrangements for eliciting and incorporating worker views and experiences
- Periodic reports of the impact that preventive schemes such as training and maintenance programs have on occupational safety and health
- Regular audits of the effectiveness of risk controls and management
- Assessments of the impact that changes such as the introduction of new work processes, procedures, or products have on safety and health
- Effective procedures for implementing new and altered legal requirements (EU–OSHA, 2012, pp. 10–14)

VISIBLE FELT LEADERSHIP

Visible felt management leadership is when managers visit the production areas and discuss safety issues and concerns with employees at their workstations and visibly show interest and concern in safety. Managers should visit the workplace as often as they can, and interact with employees at the point of action where the work is done. This is also where the risks lie, and these visits and contacts with employees will help reinforce managements' visible commitment to the safety system. These visits also indicate to the employees on the shop floor that management has not lost touch with what goes on in the workplace. Having the CEO put on work clothes and venture into the work areas also sets a good example for other managers to follow suit.

Managers can become isolated from what is happening at the workplace, and this isolation can give a false sense of security in letting the manager think that the safety system is functioning well. Doing regular safety tours also indicates to the workforce that managers are involved in the safety system. If managers are effective leaders, their subordinates will be enthusiastic about exerting effort to attain safety system objectives.

MANAGERS LOST TOUCH

One of the biggest safety complaints from employees is that their management has lost touch with what goes on at the coal face. Managers should practice "management by walkabout," and visit their respective work areas. They should speak to their employees who are vital sources of information and who are often overlooked. Employees love to talk about safety and can tell more about the current state of safety in the plant than an audit will. The scheduled site visits and safety walkabouts should be written into the key performance areas of managers at all levels as part of their safety responsibility.

POSITIVE BEHAVIOR REINFORCEMENT

Positive behavior reinforcement is another key to the success of any safety management system, and of all the functions carried out by leadership, is likely to have the most effect on the success of the system. It demands playing the ball and not the man. It requires managers to fix the problem and not try to fix the person. It forces leaders to deviate from traditional management styles when dealing with issues normally calling for disciplinary measures. It will challenge leadership at all levels, but will help create more positive leadership across the organization.

It entails recognizing workers at all levels and acknowledging that they are truly the safety and health experts in their sphere of operation. They have had more exposure to the machinery, substances, and processes in that work area than anyone else. This forms part of employee recognition. Catching people doing things right is true safety management.

WHAT IS A MANAGER?

A manager is anyone who uses management skills or holds the organizational title of "manager." A manager gets things done through other people. Management could also be the person or persons who perform the act of management and this would include any employee with responsibility over other employees or assets. Safety management consists of safety planning, organizing, leading or directing, and controlling the activities within an organization for the purpose of accomplishing safety goals and objectives. This means getting people together to accomplish desired safety goals and objectives efficiently and effectively.

BASIC SAFETY MANAGEMENT FUNCTIONS

Over the years, it has generally been accepted that a manager's main functions are:

- Planning
- Organizing
- Leading or directing
- Controlling

All these functions entail the management of employees, materials, machinery, and processes, and include the management of the safety of these assets. The four basic functions of management encompass safety management and, if integrated into a manager's normal functions, could provide for better management, leadership, and involvement.

THE PLAN, DO, CHECK, ACT METHODOLOGY

The *Plan, Do, Check, Act* (PDCA) cycle, also known as the Deming circle, is a four-step model for carrying out change and continuous improvement. Just as a circle has no end, the PDCA cycle should be repeated for continuous improvement. This iterative four-phase management method is used in business for the control and continuous improvement of processes and is recommended by most Guidelines for the implementation of safety management systems.

- *Plan*—involves establishing the objectives and processes necessary to deliver results in accordance with the expected outputs. Action plans are compiled and policies and critical safety performance indicators are established.
- *Do*—is the phase where data are collected, the plan is implemented, the safety system processes started, training commences, and standards are set.
- *Check*—is when inspections and measurements against the standard take place. This phase involves inspection, audit, measurement, and review.
- *Act*—is when corrective action is initiated to rectify deviations from planned achievements. Processes are amended, modified, and improved to create a system of continual improvement.

PRINCIPLES OF SAFETY PLANNING

Safety planning is the first step in the *Plan, Do, Check, Act* process and is what a manager does to determine action to be taken to prevent downgrading events, and to predetermine the consequences of accidents and results if such action is not taken. It encompasses the planning, design, and implementation of a safety management system (Figure 5.1).

SAFETY FORECASTING

Safety forecasting is defined as: *the activity a manager carries out to estimate the probability, frequency, and severity of loss-incurring events that may occur in a future time span.* This is aided by the risk assessment process, which is a main requirement of all safety management systems, and which is done by means of hazard identification and elimination, risk assessment, risk analysis, and risk control.

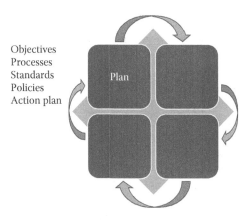

Objectives
Processes
Standards
Policies
Action plan

Plan

FIGURE 5.1   Safety planning is the first step in the Plan, Do, Check, Act process.

Safety forecasting is a planning tool that helps management in its attempts to cope with the uncertainty of the future, relying mainly on data from the past and present, and analysis of trends. It could indicate what losses could occur if a safety management system is not implemented. Safety forecasting is a decision making tool used by many businesses to help planning and estimating future risk reduction and safety system growth. In the simplest terms, safety forecasting is the attempt to predict future outcomes based on past events and management insight.

SETTING SAFETY OBJECTIVES

Setting safety objectives is when a manager determines what safety results he or she desires. This would include the action plan to implement the safety management system, the achievement of the requirements of the specific safety system Guideline, and could include achieving eventual accreditation. This would be broken down into smaller objectives, which, when achieved, will lead to attainment of the main objective. The safety management system implementation action plan would be the overall objective which would be broken down into immediate, short, medium, and long term objectives, each being time bound.

CRITICAL PERFORMANCE INDICATORS (CPIs)

These objectives are sometimes referred to as Critical Performance Indicators (CPIs), or Safety Performance Indicators (SPIs). They could be leading or lagging indicators or a combination. Traditional safety indicators were almost always lagging indicators, based on injuries and work fatalities. The obvious target would be zero (what you plan for, you normally get). Again, injuries sustained and damage caused by accidents are reactive measures and are measurements of consequence, not control. Unfortunately, they are still regarded internationally as measures of safety. A good safety management system would break from tradition and have mostly leading safety system indicators as objectives. Examples of leading monthly objectives formed into CPIs with maximum scores (for measurement) allocated are as follows:

- Safety committees formed (25)
- Near miss incidents reported (10)
- Safety observations reported (10)
- Hazards eliminated (15)
- Workplace inspections completed (5)
- Safety system audit results (5)
- Fire drills held (5)
- Number of safety toolbox talks held (10)
- Employees attending safety and health training programs or workshops (5)
- Number of safety representatives appointed and active (15)

Each objective has a rating score, and monthly management reviews of each division can be done where the achievement of objectives, and achieved scores are discussed. This is holding managers accountable for safety in the form of achievable, manageable key actions.

SETTING SAFETY POLICIES

In the *Plan, Do, Check, Act* methodology, setting safety policies is part of the planning, and is when a manager develops standing safety decisions applicable to repetitive problems, which may affect the safety of the organization. Policies and procedures concerning the safety management system would be incorporated in the form of safety system standards, procedures, checklists, and policies. Each element must have a written standard. These written standards are the measurable management performance criteria for each element of the organization's safety management system.

SAFETY PROGRAMMING

Another facet of planning in the PDCA cycle is safety programming, which is establishing the priority and following order of the safety action steps that must be taken to reach the safety objective. An example would be determining which safety management system elements are to be implemented, by when. Also important is the setting of standards for communicating this information down to the workforce. Safety management system timelines are established by safety programming and would feature prominently in the safety system action plan (Figure 5.2).

An example of an action plan objective is as follows: *To structure a 12-month action plan to implement the Safety Management System with designated authority, responsibility, and accountability with timelines for the completion of these activities.*

WHAT MUST BE DONE	BY WHO	BY WHEN	DONE
Note: CZ = Centralized DZ = Decentralized		Ongoing processes	
The new Safety Policy Statement must be reviewed, accepted and signed by all the Executives (Critical Phase)	CEO/EXCO	April	
Policy statement translated	Dave		OK
Safety and Health booklet 1.17 translated			OK
Safety and Health booklet cascaded to all areas	All Managers and DZ		
Policy Statement framed and displayed			
Policy Statement circulated	Pub. Dept.	May	
Policy included in all documents Website Induction training Contracts	Jim	May–ongoing	
The ARCFLASH project to receive priority. Suggest an outside specialized company to do project	WB	June–Sept.	
Standard Element 1.1 Safety Policy written Approved Cascaded Website	JTR	April	

FIGURE 5.2   An example of a safety management system action plan.

SAFETY SCHEDULING

Safety scheduling is when a manager establishes time frames for the implementation of safety management system elements. When introducing and maintaining a safety management system, a schedule would be determined for the introduction phase of the system, the training phase, as well as the follow up and review of the results, successes, and failures of the system.

Time Span

Many undertakings underestimate the amount of time needed to fully implement a structured and functional safety management system. The time required would differ from business to business depending on the status, content, and effectiveness of their existing safety management efforts. When starting from ground level, an organization should expect the full implementation to take up to 5 years. This also varies from industry to industry and country to country. Organizations that already have a working safety system in place may be able to accomplish the implementation in a shorter period. The safety scheduling function of management will determine the time line needed to draw up the action plan accordingly.

SAFETY BUDGETING

Safety budgeting means allocating financial and other resources necessary to achieve the safety objectives. A budget allocation may be required for the safety management system implementation and maintenance. Funds should be allocated for repairs, safety equipment, training, etc. Safety promotion and publicity should also be budgeted for. Mechanical or structural repairs or modifications may be needed to eliminate hazards reported through the system and these expenses must also be budgeted for. High-risk situations should enjoy priority and budgets must allow for their immediate rectification.

ESTABLISHING SAFETY PROCEDURES

Procedures are required for a number of different operations in industry and mines. These procedures help identify potential areas of accidental loss, which will be eliminated by following the procedure correctly.

Establishing correct safety procedures is a management function that analyzes certain tasks and ensures that safe work procedures (SWPs) are drafted for performing the work. These procedures are sometimes referred to as a Job Safety Analysis (JSAs), Critical Task Procedures (CTPs), or Safe Work Procedures (SWPs). They should be derived by means of an analysis process such as the widely used Job Safety Analysis (JSA) procedure, or something similar. Based on the analysis, the jobs can be risk ranked and the critical tasks identified. This will help prioritize the writing of procedures.

Many safety systems make the mistake of insisting that there must be a written procedure for every job in the organization. A lot of time and effort can be wasted by doing this. By applying the management principle of the *Critical Few*, procedures should ideally be written for those tasks or jobs that have a higher risk than others based on the task risk analysis exercise.

SAFETY ORGANIZING

Safety organizing is the "Do" function in the *Plan, Do, Check, Act* sequence and is the function a manager carries out to arrange safety work to be done most effectively by the right people. It involves integrating safety into the organizational structure. Safety organizing entails the matching of the organizational form, such as structure, reporting relationships, and information technology, with the organization's strategy.

Safety organizing means bringing together physical, human, and financial resources to achieve safety objectives. Management then identifies activities to be accomplished, classifies safety activities, assigns activities to groups or individuals, then delegates authority and creates responsibility (Figure 5.3).

APPOINTING EMPLOYEES

Appointing employees is a management function where management ensures that the person is both mentally and physically capable of carrying out the work for that position, safely. One of the elements of the safety management system would be the criteria for the selection and placement of the correct person in the right job.

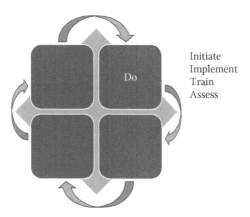

Initiate
Implement
Train
Assess

FIGURE 5.3 Safety organizing forms part of the *Do* in the PDCA sequence.

Safety Department

The function of all safety departments and personnel should be to guide, educate, train and motivate all levels of management, workers, and unions in the techniques of accident and disease prevention and to advise on and coordinate the safety management system (GETMAC). Theirs should be a staff function and not a line function.

When appointing safety and health department employees, management has a duty to the profession to ensure suitably qualified and experienced practitioners are selected. Proper job descriptions based on the American National Standards Institute (ANSI) guidelines: ANSI/ASSE Z590.2–2003 *Criteria for Establishing the Scope and Functions of the Professional Safety Positions* should be used as selection and training criteria for safety staff. Similar standards are available for other professions as well.

The safety and health personnel are vital employees with regard to the coordination of the system and their selection must be carefully considered. Management's function is to appoint the right people in the right positions. This would also encompass allocating persons to coordinate the safety management system implementation and maintenance. Safety organizing would entail the appointment of the following:

- The Safety Department
- Safety System Coordinators
- Industrial Hygienist
- Accident Investigators
- Safety Representatives
- Safety Committee Chairperson
- Management Safety Champion
- Safety Trainers
- Internal Auditors, etc

Developing Employees

A manager develops employees by helping them improve their safety knowledge, skills, and attitudes. Management has to ensure the safety and health staff are up to date with the latest trends in safety and risk management and that there is an ongoing self-development program in place for them. Further studies, as well as membership in local, regional, and national safety and health associations, should also be encouraged.

SYSTEM INTEGRATION

The safety management system should be integrated into the day-to-day business of the organization and should not be regarded as a stand-alone item divorced from normal operations. The more the safety management system is integrated into the organization's operations, the more effective it will be.

SAFETY DELEGATION

Safety delegation is what a manager does to give safety authority and entrust safety responsibility to his or her subordinates, while at the same time creating accountability for safety activities. All employees should be given the authority to report unsafe or unhealthy situations. Once hazards have been identified, managers are then held accountable to rectify the deviations depending on their area of control and level of authority and responsibility.

The key to the success of a safety management system is allocation of authority, the creation of safety responsibility, and the holding of managers and employees at all levels accountable for safety actions in the form of doing what the safety system standards require of them. Traditionally, individuals were only held accountable after an accident and this accountability was in the form of allocating blame.

CREATING SAFETY RELATIONSHIPS

Creating safety relationships is work done by a manager to ensure that safety work is carried out by the team with utmost cooperation and interaction among team members. The safety management system must be owned by all levels in the organization and should not be seen as a bargaining tool, or a system to gain personal benefits or demands. The system requires participation, support, and action from all levels within the organization and cannot be left as one person's, one department's, or one manager's responsibility.

SAFETY AUTHORITY

Safety authority is the total influence, rights and ability of the position, to command and demand safety. Management has ultimate safety authority, therefore it is the only level that can effectively implement and maintain an effective safety management system. Leadership has the authority to demand the implementation of safety system elements and also the authority to take necessary remedial actions to ensure that standards, policies, and procedures are implemented and maintained.

Safety Responsibility

The safety management system Guidelines emphasize that occupational safety and health is the responsibility of management, starting with the most senior managers. This cannot be over emphasized as many still think that safety is the sole responsibility of the safety manager and the safety department. This is one of the biggest safety paradigms that need to be changed.

Safety responsibility is the safety function allocated to a post. It is the duty and function demanded by the position within the organization. This duty lies with all levels of management as well as with employees. The higher the management position, the higher the degree of safety authority, responsibility, and accountability. One cannot be held accountable for something over which one has no authority. The degree of safety accountability is proportionate to the degree of safety authority. Job descriptions are vital management tools and should clearly define the safety authority, responsibility, and accountability for all levels within the organization. The safety management system's safety standard must clearly define these relationships for the system to be a success.

Safety Accountability

Safety accountability is when a manager is under obligation to ensure that safety responsibility and authority is used to achieve both safety system and legal safety standards. Employees also have safety accountabilities but in proportion to their safety authority.

Leadership has the accountability to manage the safety management system and to provide the necessary infrastructure and training to enable the system to work. Employees should be held accountable for participating in the system and following company safety policies, procedures, and practices.

Management at all levels is then held accountable to rectify the problems identified by the ongoing risk assessment process, as well as the management review and safety system audits, to ensure that the high-risk acts or conditions highlighted by these systems are rectified and do not recur.

Senior management must set standards for the physical work environment as well as standards of accountability for all levels. These standards are then measurable management performances. An example of a letter of appointment, and acceptance, as required by a safety management system is as follows:

In Terms of the Organization's Safety Management System (Element #1)

All Managers, Supervisors, and Contractors (including acting managers)

As per the organization's safety management system, Element (#1): You are hereby appointed responsible and accountable for the safety, health, and well-being of all employees, including contractors, under your responsibility. This appointment involves accepting safety and health responsibility and accountability, as well as for the implementation of all the safety and health management system elements and other safety requirements listed in the organization's safety manual or other general instructions, in your areas of responsibility.

Note: Before signing, please note that I hold you personally responsible for the safety and health of your employees and will offer any support or assistance I can to help you maintain a safe and healthy work environment and safe work habits among your employees.

Kindly accept this appointment by signing in the appropriate place and by returning the original to me.

Yours sincerely,
Organization CEO

An example of the written acceptance of this responsibility, which is expected from all managers appointed is as follows:

I . . . have read and understood the organization's safety management system standard (Element #1) *Managers Designated Responsible for Safety and Health.* I accept its content and accept that I am responsible for the safety of my employees/contractors and for implementing and maintaining the Organization's Safety System. This also applies to my acting managers.

THE FUNCTIONS OF DIRECTING (LEADING) SAFETY MANAGEMENT SYSTEMS

SAFETY LEADING

Safety leading is what a manager does to ensure that employees at all levels act and work in a safe manner and that the necessary controls, in the form of safety management system processes, are in place to ensure a safe working environment. Leading is connecting with employees on an interpersonal level. This goes beyond simply managing tasks and involves communicating, motivating, inspiring, and encouraging employees toward a higher level of productivity. It entails taking the lead in safety matters, making safety decisions, and always setting the safety example. This is one of the most important management functions in implementing and maintaining a safety management system.

CREATING TEAMS

Management would create teams to address certain safety projects as part of management leading. These teams would then tackle projects that need solutions in order to enhance the safety management system. A team is often formed to write the safety system element standards, or to focus on specialized elements such as lockout, tag-out, try out procedures, and other processes. Other teams could be used to develop safety competition criteria, another to produce the safety website or newsletter. Creating teams to work on safety system implementation uses the principle of involvement, which states that the more employees are involved in the safety system, the more they will be motivated to participate.

INTEREST IN SAFETY

The principle of interest in safety, is that the workforce will only become interested in safety if management shows an interest in safety, and the results achieved by them individually and as a group. Management must set the safety trend and help them achieve safety objectives.

SAFETY ROLE MODEL

Top management must take the lead in a safety management system. The chief executive must lead the process by chairing the executive safety committee meetings and by ensuring all appointments are done and kept up to date. The subordinate managers then chair the lower level committees, and this process is cascaded down to all levels of supervision. The Chief Executive Officer (CEO) is a role model for all other managers in the organization. An employee will voluntarily follow the directions of a leader because they believe in who he or she is as a person, what they stand for, and for the manner in which they are inspired. Only managers have the authority and ability to create and maintain a safe and healthy workplace, and it is their duty to do so.

MAKING SAFETY DECISIONS

Making safety decisions is when managers make a decision based on the facts presented to them. These decisions include the safety management system implementation action plan, system standards, the timeline, as well as the critical safety performance indicators and many other facts.

SAFETY COMMUNICATION

Safety communicating is what a manager does to give and get understanding on safety matters. Management and employees' expectations concerning participation in the safety management system must be clearly communicated. Standards must be set and communicated to all concerning the requirements of their role in the safety process.

One reason for the failure of safety management systems is that the communication around the system and its elements is poor. The more employees are informed about safety issues affecting them, the more they are likely to participate in those activities. A structured system entails two-way communication and total transparency. The elements of the safety system that aid communication are as follows:

- Safety committees, including minutes
- Safety newsletters
- Reports, annual and audit
- Accident investigation results
- Safety website updates
- Toolbox talks

- Safety induction and refresher training
- Employee safety handbook
- Safety system element standards
- Safety bulletin boards, etc.

MOTIVATION

Motivating for safety is the function a manager performs to lead, encourage, and enthuse employees to take action and participate in the safety management process. Positive participation in the safety management system should be acknowledged and rewarded. Safety motivation increases as people are given recognition for their contribution to the safety effort. Commending and encouraging employees for safety actions go far in ensuring that those actions are repeated. Good safety should be praised, and this praise should be made public where possible. The more employees are recognized for being a part of the safety drive, the more they will be motivated.

CANE OR CARROT?

Safety has traditionally been a system of punishing employees for safety violations or rewarding them for being injury free (the cane or carrot method). Both these approaches have made a mockery out of good safety intentions, as employees hide unsafe behavior and do not report injuries, and for this they are rewarded. Modern safety management views employees as team members and active participants in the safety system. Injuries indicate a failure in the system and the aim of investigations should be to rectify the breakdowns and not punish the employee. Discipline is always the last resort in solving safety breakdowns and positive discipline should receive preference.

FOCUS ON THE SYSTEM

A formal safety management system is driven by audit and measurement of the ongoing processes in place to reduce risk and eliminate undesired events. The system is not gauged by the number of injuries experienced and this, therefore, changes the paradigm that injuries are the only measure of safety performance. The international Guidelines now bring about a major change in how safety is perceived and measured. They now call for organizations to focus on the safety effort rather than the injury experience. This moves the safety focus from the employee to the organization and its systems.

SAFETY CONTROLLING

Safety controlling is the management function of identifying the risks, identifying what must be done for safety, inspecting to verify completion of work, evaluating performance, and following up with safety action and commendation. This is the most important safety management function, which is an expanded version of the *Plan, Do, Check, Act* sequence (Figure 5.4).

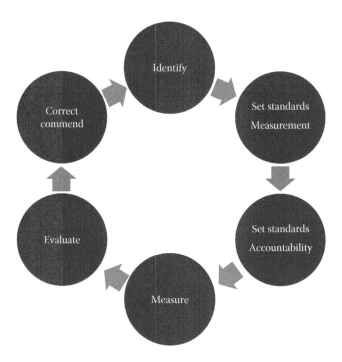

FIGURE 5.4 Identify, Set Standards, Measure, Evaluate, Correct, Commend (ISSMECC).

IDENTIFICATION OF THE RISK, AND SAFETY MANAGEMENT WORK TO BE DONE

Based on risk assessments, a manager lists and schedules the work needed to initiate, implement, and maintain a safety management system in order to create a safe and healthy work environment. This would mean the introduction of a suitable structured safety management system based on world's best practice. All safety management systems should be based on the nature of the business and be risk-based, management-led, and audit-driven.

SAFETY MANAGEMENT SYSTEM (SMS)

A *safety management system* is defined as: *ongoing activities and efforts directed to control accidental losses by monitoring critical safety elements on a continuous basis.* Monitoring includes the promotion, improvement, and auditing of safety management system elements regularly.

PRINCIPLE OF THE CRITICAL FEW

The principle of the critical few states that, "*A small number of basic causes could give rise to the majority of safety problems.*" A few critical jobs, or procedures, could be responsible for the majority of injury causing accidents occurring, so these few critical items (critical safety elements) should receive maximum safety control to minimize their potential for causing (the majority) of problems.

Precontact control is directing the safety efforts toward controlling these crucial areas before a loss occurs. Many safety systems are reactive and only institute controls after a loss has occurred. This is termed post-contact control, firefighting, patch prevention, or treating the symptom and not the cause.

SET STANDARDS OF PERFORMANCE MEASUREMENT

Safety standards are measurable management performances. Standards are set for work to be done to maintain a safe and healthy environment, free from actual and potential accidental loss. A safety standard of performance is a management approved expression of the performance, requirements and expectations that must be met, and which can be measured against.

Safety system standards should be objective, measurable, realistic, and clearly stated. The standards should be written in terms of specific measures that will be used to gauge performance.

Standards should be established in writing for all the safety management system elements. Without standards, the management system has no direction, nor are safety expectations established. (If you don't know where you're going, any road will take you there.)

The safety management system standards for each element and sub-element of the system must indicate who does what, and when it is to be done. It must include measurable criteria. These standards must be issued by senior management and not the safety department. They should be agreed to by management and the workforce and should be reviewed at least annually or more often if changes occur.

An example extract from a safety system standard for Occupational Safety and Health Committees is as follows:

Objective

- To establish and sustain a system of Occupational Safety and Health (OS&H) Committees throughout the organization to facilitate the management of OS&H issues and to provide a forum for OS&H communication, the organization safety system implementation, and decision making
- These committees can recommend policy changes, but they are not accountable for OS&H policies of the organization. They do not replace the single point accountabilities for the organization's safety system, which remains with line management

Standard

- OS&H committees are to be established at four different levels and in different departments throughout the organization for the purpose of providing forums for communication, decision making, and the organization safety system implementation

- Each OS&H committee will meet on a regular monthly basis and these meetings will be scheduled in advance to ensure attendance is given priority
- Meetings should generally be limited to 1 hour in duration
- The secretary must submit a pre-prepared agenda of items for discussion to the committee members, 1 week before the actual meeting
- Members are to submit discussion points to the Committee Chairperson at least 1 day before the meeting
- Brief minutes of the meeting, noting the decisions, actions, and assigned responsibilities shall be taken and circulated within 3 days of the meeting by the secretary of the committee. A copy is sent to the next higher-level Committee Chairperson
- Minutes will be in the form of an action plan listing:
  - WHAT must be done
  - WHO will do it
  - WHEN it shall be done by
- Where possible, minutes of the OS&H meeting should not be longer than 1 page. The minutes are to be issued within 3 days of the meeting

SET STANDARDS OF SAFETY AUTHORITY, RESPONSIBILITY, AND ACCOUNTABILITY

This is perhaps the most important step in implementing a safety management system and is proposed by most of the system Guidelines. Management set standards of accountability by delegating authority to certain positions for ongoing safety management system work to be done. Coordination and management of the safety management system needs to be allocated to certain departments and individuals, and this standard dictates who must do what, and by when.

Senior Management Appointment

Appointments for safety authority, responsibility, and accountability should be made for all levels within the organization, including employees. Previously, a general belief that "Everybody is responsible for safety," led to a situation where no one was responsible when things went wrong. Setting standards of authority, responsibility, and accountability makes it clear as to exactly what an individual's safety responsibility is.

Some Guidelines recommend that the CEO appoint a manager at a senior level to be responsible for the implementation and maintenance of the safety system and that this person's appointment be made known to the entire organization. The appointee would be responsible for reporting progress to the executive management on a regular basis. These appointments must be done in writing, spelling out the responsibilities, and accountabilities of the appointments. Other levels of management and employees at all levels within the organization should also be appointed for safety responsibility within their span of control, and within the bounds of their authority.

The following example of a safety system standard spells out safety responsibilities for all levels within the organization:

1. Objective
   a. To define the safety and health responsibilities, accountabilities and authority to act from the CEO, Executive, and Vice Presidents to all subordinate line managers and employees
   b. Further, to make all levels of management responsible and accountable for the implementation of the safety and health management system in their areas of responsibility
2. Definitions
   a. Safety Authority: Safety authority is the right or power assigned to an executive or a manager in order to achieve certain organizational safety goals
   b. Safety Responsibility: Safety Responsibility is a *duty, obligation,* or *liability* toward safety for which someone is accountable
   c. Safety Accountability: Safety Accountability is being liable to be called on to render an account on safety and be answerable for safety issues
3. Responsibility
   a. The CEO: The CEO is responsible for the issuance of the occupational health and safety policy statement and will demonstrate and accept safety accountability by:
      i. Ensuring that there is an appropriate occupational safety and health management system in place and functioning at the company
      ii. Holding regular safety executive (EXCO) team meetings
      iii. Providing resources and support to achieve the occupational safety and health management system objective plan
   b. The Executive Vice Presidents, Vice Presidents, Executive Directors, Managers, Supervisors, and Contractors must:
      i. Implement the company occupational safety and health management system in their area (s) of responsibility, in a manner to ensure that the workplace hazards and other risks are reduced to an acceptable level
      ii. Ensure that the company occupational safety and health management system standards are implemented, so that accidents resulting in injuries, occupational diseases, and property damage or business interruption are minimized or eliminated
      iii. Providing resources and support to achieve the occupational safety and health management system objective plan
      iv. Monitor the Safety Performance Indicators (SPIs)
   Each Executive Vice President, Vice Presidents, Managers, Supervisor, and Contractors should be an active participant in the company occupational safety and health management system
   c. Employees must:
      i. Follow the standards of the safety and health management system and learn about the latest approved updates and standards

    ii. Contribute to the implementation of the safety management system by reporting hazards, accidents, injuries, and near miss incidents

    iii. Commit to safe work procedures and the approved PPE

4. Authority

  a. The CEO at the company has the authority to:

    i. Adjust permanent manning levels to ensure safe operations and to approve the company safety system for implementation

    ii. Approve and take decisions concerning the implementation of the company safety system

    iii. Provide resources and support to achieve the occupational safety and health management system objective plan

    iv. Decide on all matters relating to safety

    v. Approve safety policy, standards, and instructions

    vi. Issue safety executive orders

  b. The Executive Vice Presidents, Vice Presidents, and Executive Directors at the company have the authority to:

    i. Review and set priorities of any safety related work

    ii. Request the necessary resources from within the organization to control any unsafe condition or behavior

    iii. Implement and maintain the elements of the company occupational safety and health management system

    iv. Stop production under their control where safety is compromised

    v. Commend any staff reporting to them on safety related issues where needed

    vi. Finalize formal warning/disciplinary proceedings for any staff or contract employees reporting to them for safety relating issues leading to dismissal. These proceedings must be in accordance with, and cannot exceed, the company policies and disciplinary measures. In this regard, all labor law rules and regulations will be adhered to

  c. Managers and Supervisors (including Contractors) have the authority to:

    i. Direct and instruct non-supervisory employees, and peers if necessary, on safety related issues on behalf of the Management

    ii. Implement and maintain the elements of the company occupational safety and health management system

    iii. Stop production under his/her control where safety is compromised

    iv. Request the necessary resources from within the organization to control any unsafe condition or behavior. Request the necessary resources from outside the organization to control any unsafe condition

    v. Initiate formal warning/proceedings for supervisory staff for safety related issues leading to warnings in accordance with company policies labor law, rules, and regulations

    vi. Finalize formal warning or discipline proceedings for supervisory and contract employees/staff reporting to them for safety related issues leading to suspension

       vii. Commend any staff reporting to them on safety related issues where needed

   d. Everyone at the company (including contractors) has the authority to:

       i. Stop their job or task to ensure safety

       ii. Report unsafe situations to their immediate supervisor

5. Accountability

Safety accountability therefore designates management at all levels within the company as responsible and accountable for safety and health of their respective areas, and within their levels of authority

All employees are responsible and accountable for conducting their work in compliance with established occupational safety and health standards as dictated by the company's occupational safety and health management system, Job Safe Practices (JSPs), procedures, permits, precautions, laws, and regulations

   a. The CEO at the company:

   Will demonstrate and will accept safety accountability by:

       i. Ensuring that the company safety system is in place and functioning

       ii. Holding and leading the monthly EXCO meetings

       iii. Setting an example to all echelons of management by acting safely

       iv. Participating in safety functions and meetings

       v. Reporting to the Board of Directors on the achievement of objectives within the occupational safety and health management system action plan

   b. Executive Vice Presidents, Vice Presidents, and Executive Directors:

   Will demonstrate and will accept safety accountability by:

       i. Setting and communicating clear safety objectives within the company occupational safety and health management system for his/her business line

       ii. Ensuring that the company occupational safety and health management system is effectively implemented in his/her business line

       iii. Ensuring that accident investigations are undertaken and completed promptly and comprehensively, and that immediate and root causes are identified and eliminated or mitigated within an established time frame and follow up implementation

       iv. Not giving an instruction to any member of his/her employees to perform a high-risk act

       v. Ensuring the development and implementation of appropriate safety standards for their area of responsibility

       vi. Ensuring compliance with all Job Safety Practices (JSPs) and safety standards

       vii. Taking immediate action to eliminate or control any unsafe condition or behavior that exists or is brought to his/her attention in his/her area of responsibility

       viii. Ensuring that managers approve all JSPs relevant to their area

       ix. Ensuring that the JSPs and safety standards relevant to his/her area are fully comprehensive and up to date

    x. Chairing the area monthly safety committee meetings

    xi. Setting an example to all under his/her control by acting safely

c. Managers and Supervisors (including Contractors) at the company will demonstrate and will accept safety accountability by:

    i. Setting and communicating clear safety objectives within the company occupational safety and health management system for his/her department

    ii. Taking ownership of, and applying the occupational safety and health management system standards applicable to his/her work area and employees

    iii. Ensuring that accident investigations are undertaken and completed promptly and comprehensively, and that immediate and root causes are identified and eliminated or mitigated within an established time frame and follow up implementation

    iv. Not giving an instruction to any member of his/her team to perform a high-risk act

    v. Ensuring compliance with all JSPs and safety standards relevant to the team

    vi. Taking immediate action to eliminate or control any unsafe condition or behavior that exists or is brought to his/her attention in their area (s) of responsibility

    vii. Approving all JSPs relevant to his/her area of control

    viii. Ensuring that the JSPs and safety standards relevant to his/her area of control are fully comprehensive and up to date

    ix. Holding monthly safety committee meetings with supervisory staff and contractors in their area (s) of responsibility

    x. Setting an example by acting safely

d. Everyone (including Contractors) at the company

Will demonstrate and will accept safety accountability by:

    i. Taking accountability for their own safety behavior

    ii. Not accepting an instruction to perform a high-risk act

    iii. Reporting accidents and injuries/illnesses, near miss incidents, and unsafe situations or behavior to his/her immediate supervisor

    iv. Rectifying unsafe situations within his/her area of authority

    v. Complying fully with all JSP's and safety standards in the performance of their work

    vi. Obeying the company driving rules at all times

    vii. Participating in the area monthly safety committee meetings

    viii. Setting an example to all by acting safely

MEASUREMENT AGAINST THE STANDARD

A vital component in the safety control process is measuring against the standards that have been set. This is the "Check" component in the *Plan, Do, Check, Act* methodology (Figure 5.5).

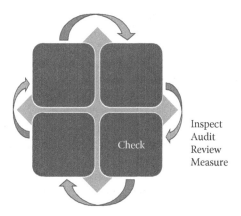

Inspect
Audit
Review
Measure

FIGURE 5.5 The *Check* segment of the PDCA cycle.

By carrying out safety inspections, the condition of the workplace and the ongoing processes are now measured against the accepted safety management system standards. Physical inspection of the workplace will highlight deviations from standards and indicate whether the safety system is working. Measurement of the system determines if the requirements of the written standards are being applied at the workplace.

Measurement against the standards is carried out in a number of ways required by the safety management system. Some examples are: formal site inspections, Safety Representative inspections, management tours, informal inspections, internal and external audits, management reviews and documented evidence review, inter alia.

Some of the measurements could be the following:

- Is the business order (housekeeping) up to standard?
- Are safety signs and notices displayed and current?
- Are all machines and sources of energy guarded?
- Work areas demarcated?
- Hazardous chemical storage in order?
- Have areas been inspected?
- Wearing of PPE?
- Lockout, tag-out standard being applied?
- Work permits displayed and current?
- Evacuation drills held?
- Fire equipment unobstructed?
- Notice boards up to date?

What gets measured gets management's attention, and if there is no formal system of measurement, then management does not know how well the safety management system is doing compared to its own standards and best practice.

EVALUATION OF CONFORMANCE

Depending on which measurement method is used, the results are now quantified in the form of a safety system audit during which a percentage is allocated, marks given, or a ranking against established safety system standards, is determined. Safety audits, both internal and external, evaluate compliance with an organization's standards, and the final scores then indicate where there is a deviation from the standards.

CONTINUAL IMPROVEMENT CYCLE

Safety system audits are a valuable management measurement tool and will indicate the strengths and weaknesses of the system. An audit should take place every 6 months by a selected team of trained internal safety system auditors. External audits by third party auditors will give management an accurate snapshot of the effectiveness of the safety system, and will also allocate a measurement in the form of a score for each element. This will provide a basis for continual system improvement (Figure 5.6).

CORRECTIVE ACTION

Taking corrective action is the "Act" phase of the *Plan, Do, Check, Act* methodology. It involves taking corrective action on deviations, amending procedures or processes, modifying where necessary, and improving weaknesses. The amount of corrective action will be proportional to the amount of deviation from standards set. Corrective action may involve enforcing the safety system standards and taking the necessary action to regulate and improve the methods.

Corrective actions are where the deviations from standards are rectified. A deviation indicates that a risk has not been controlled or a process is not working correctly. This may mean fixing the high-risk condition or correcting high-risk behavior, or a combination of both, and then identifying and rectifying their root causes.

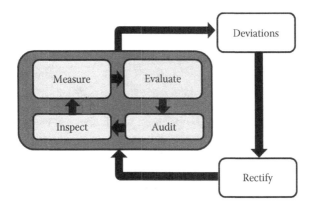

FIGURE 5.6   The continuous improvement cycle.

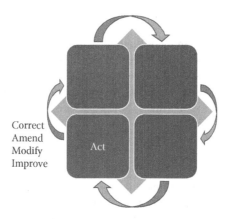

FIGURE 5.7   The *Act* phase of PDCA methodology.

Once again, standards are established for these corrective actions which state who-must-do-what and by when it must be done, in order to get the situation rectified. Corrective action must be positive, time related, and be assigned to responsible people (Figure 5.7).

COMMENDATION

Commendation is when a manager pays compliment and expresses gratitude for adherence to an achievement of pre-set safety standards or requirements. This is applicable across the board and is a vital component of the safety management system. If employees are not recognized for participating in the system, their enthusiasm will soon wane. If the safety management system is introduced as another, "flavor of the month," employees will not integrate it into their daily routines, but rather treat it as another safety fad that will soon fade.

Management's involvement, support, and participation in the safety management system are vital. This integration of the system into the day-to-day activities of employees at all levels will cause the safety management system to improve the safety culture of the organization.

SAFE BEHAVIOR RECOGNITION

The safety management system should provide mechanisms for recognizing positive and safe behavior as well as safe work areas, which conform to the safety requirements of the system. A part of the recognition system should be the acknowledgement of safe behaviors and safe work conditions. Often, management tends to focus on the negative aspects of safety, yet a powerful tool is to recognize the positive. In safety, positive behavior reinforcement goes a long way in achieving employee participation and involvement in the safety management system. Employees appreciate recognition and feel motivated to participate in safety activities. Recognition will enthuse and encourage workers to support the system and lead to its success.

SUMMARY

Safety management control eliminates the root causes of accidents by setting up a series of safety processes and by delegating safety responsibility and accountability. This system creates a work environment in which high-risk acts and conditions, and possible loss producing events, are reduced.

6 Audit-driven Safety Management Systems

INTRODUCTION

The *Plan, Do, Check, Act* methodology calls for the safety management system (SMS) to be reviewed after implementation. The safety management system audit is part of the "Check" sequence (Figure 6.1).

SAFE WORK ENVIRONMENT INDICATORS

An absence of, or a low number of serious injuries, does not necessarily indicate a safe work environment. This is a negative, downstream measure largely dependent on luck. A safe work environment should be the result of active participation by management and staff in identifying risks and then mitigating them. The identification and prompt elimination of hazards and risks is a safety management system's approach that needs to be quantified by audit.

SAFETY SYSTEM REVIEW

Safety system reviews are part of the "Check" stage in the *Plan, Do, Check, Act* methodology and a major component of management controlling, that is, evaluating performance. Safety audits measure the management work being done to reduce risks and control losses, and are, therefore, vital performance indicators for the safety system.

A management review of the safety management system is a check to see where the system is working and where it is not working. Safety inspections are the measurement tool and a review would mean a complete audit of the safety management system entailing a comprehensive inspection, employee interviews, and documented evidence scrutiny.

SAFETY SYSTEM AUDITS

DEFINITIONS

- A safety system audit provides the means for a systematic analysis of each element of a safety and health management system to determine the extent and quality of the controls.
- A safety audit is a critical examination of all, or part, of a total safety and health management system.
- A safety audit is a management tool that measures the overall operating effectiveness of a company's safety and health management system.

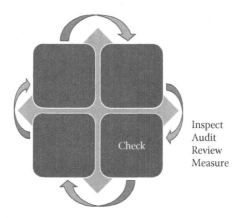

Inspect
Audit
Review
Measure

FIGURE 6.1   The *Check* stage of the PDCA methodology.

NOT AN INSPECTION

It must be emphasized that an audit is not simply an inspection of a workplace. Many safety professionals incorrectly call a safety inspection a "safety audit," which it is not. A safety inspection entails inspecting the work areas and checking against a checklist or identifying hazards and noting them down on an inspection checklist. This is *not* an audit.

An audit involves an opening conference, an inspection, a document review session, a scoring session, and a close out conference, followed by a comprehensive audit report indicating element and total scores.

REASONS FOR AUDIT

Past experience has shown that management always believes that their safety performance is much better than it actually is. Auditing the safety and health system is the only positive and progressive method of measuring safety performance, and of rectifying managements' misconceptions of how well they are doing in safety.

REACTIVE MEASUREMENT

Most organizations only measure safety performance by the number of disabling or lost time injuries that they experience. This performance is traditionally presented as a disabling injury incidence rate (DIIR) or disabling injury severity rate (DISR). Only a few accidental events result in major injuries. Less than 2% produce minor injuries, and about 4% cause property damage. Since the majority of accidental events do not result in injuries, using their data as a sole safety gauge does not show the complete picture. Injury rates are lagging indicators and only measure the consequences of poor control in the form of safety system failures. As they are measurements of consequence, largely dependent on fortuity, they do not accurately reflect the work being done to control accidental losses. Safety audits measure both upstream and downstream indicators of a safety system.

More meaningful information about the safety system is obtained from systematic inspection, auditing of physical safe guards, systems of work, rules and procedures, and training methods, than on data about injury experience alone. The information on injury experience is reactive and not control. Audits measure the amount of control an organization has over its risks.

MEASUREMENT AGAINST STANDARDS

Management is familiar with measurements and quantified information. By putting a measurement on something is the same as getting it done. By quantifying safety, the organization can focus on its safety weaknesses and build on its safety strengths. Only by identifying deviations by means of audits, can a company rectify its safety system shortcomings. Measurement is an ongoing process of comparison with standards. Without adequate standards, there can be no meaningful measurement or evaluation of safety systems.

BENEFITS OF AUDITS

An audit of the safety system measures the system's strengths and weaknesses and the degree of conformance to the system requirements, standards, procedures, and processes. The audit compares work being done, and standards being maintained, with accepted safety and health standards, the organization's standards, and legal requirements. A safety and health audit will help prioritize work to be done to further reduce the risks of the business. A systematic and thorough audit will initiate the development of a viable improvement plan, which can then be implemented and communicated to all levels.

Audits help review the safety system activities during the previous period, and provide information for safety system continuous improvement. Audits highlight safety system strengths, weaknesses, opportunities and threats, and are a learning experience for the organization and its employees.

RECOGNITION

Audits also give recognition. Aspects of the safety system that are running well are identified and recognition is given for meeting or exceeding these standards. Audits also help to refocus management's attention on specific safety system elements.

ACCIDENT ROOT CAUSES

The principle of safety definition states that decisions concerning the safety system can only be made if the root causes of loss producing events are clearly identified. A safety system audit helps identify these root causes and directs the safety efforts to eliminate them. These then become the new objectives to be achieved to improve the safety system.

LEGAL COMPLIANCE AUDIT

Complying with the local safety and health regulatory requirements and laws is of paramount importance, and this compliance should form the foundation of the safety management system. Many of the legal elements are basic safety and health processes designed to protect the safety, health, and welfare of employees. The safety and health policy must have a reference to this compliance, and it should be seen as the minimum to strive for. Safety systems must build on these legal requirements and exceed them wherever possible.

INTERNATIONAL COMPANIES

Organizations that operate international divisions are under obligation to conform to the host country's safety and health laws as well. Many choose to maintain conformance to and compliance with the parent country's safety laws, if they exceed the local requirements. A good safety management system should exceed both requirements and regard the safety labor laws as the minimum to strive for.

AUDIT FREQUENCY

Since an audit measures and quantifies the effectiveness of the safety management system and conformance with the company's safety and health standards, internal audits of the entire safety system should be carried out every 6 months and external audits, annually. The work done during the period preceding the audit is what is audited and measured. Credit and scores cannot be given for work planned for the future. Efforts, achievements, and safety system activities during the previous 12 months are what should be audited and quantified. Since most standards within the safety system are time related, it makes sense to audit what took place during a specific period of time, and a 12-month period is ideal for external, or third party, audits. The periods are recommended so that a substantial history of safety activities can be accumulated and allows for the maturity of the system. More frequent audits can be done dependent on the organization's needs.

EXAMPLE

The organization's standard calls for one safety inspection per month to be carried out in each department. This means that evidence should be presented which shows that each department carried out at least twelve inspections in the 12 months preceding the audit.

AUDITABLE UNITS

Before any safety system audit, it should be determined if the organization represents an auditable unit. Ideally, the unit to be audited should be geographically, organizationally, and operationally together on one property and be substantial enough to maintain a comprehensive 70-element (or more) safety system.

Audits of small off-site depots, which are unmanned most of the time, make it hard to audit as separate units, as many elements of a safety management system are not applicable to these depots. Small warehouses with two or three employees are also regarded as non-auditable units, and while they need to be inspected and conform to the safety system, they are not suitable for fully fledged audits. Should these smaller, outlying depots and warehouses form part of the larger organization, they should be inspected and included in the complete audit of the organization and not be audited separately.

TYPES OF AUDITS

There are various types of audits and examples are as follows:

- Baseline audit
- Benchmarking
- External third party audit
- Internal self-audit
- Informal audit
- Formal audit

BASELINE AUDIT

A baseline audit is an audit of the entire safety management system to establish the status of an organization's safety management system in comparison to world's best practices. This is normally the first formal, thorough safety audit that the organization has been subjected to, and will deliver an accurate snapshot as to what is in place and working, what is missing, where deviations are, and exactly where the safety management system is at that point in time.

BENCHMARKING AUDIT

Safety benchmarking is the process of comparing one organization's safety management processes and performance metrics to the best in industry or best ranked safety companies. Properly executed, the audit must measure the gap between the organization's safety system and safety systems operating in similar category organizations. Benchmarking by comparing injury and fatality rates only, is neither accurate nor meaningful.

The process of best practice benchmarking of safety entails management identifying the best organizations in their class of industry, and comparing the results and processes of those being studied to one's own results and processes. This enables the organization to learn how successful companies manage safety and what safety system and safety processes they use to make them successful.

Benchmarking is used to measure performance using specific indicators, and in safety, the injury rates should not be the only metrics used. In modern safety management, upstream safety activities are far better benchmarks. These are leading indicators over which management has direct control, while injury rates are lagging indicators which can be prone to under reporting and manipulation. The Luck Factors that influence the outcomes of many accidents also lead to inaccurate comparisons if only the injury rate is used.

Upstream comparisons could include the following:

- Number of employees attended formal safety training
- Number of hazards reported
- Near miss incidents reported
- Number and frequency of safety committee meetings
- Written safety system standards reviewed
- Internal audit scores
- Safety system standards updated
- Number of fire drills held
- Management tours conducted
- Toolbox talks held, etc.

Downstream measurements could include the following:

- Lost time injury rate
- Restricted duty injury rate
- Total injury rate
- Fatality rate
- Injury severity rate
- Property damage costs
- Vehicle damage costs, etc

EXTERNAL THIRD PARTY AUDIT

The safety Guidelines are ideal benchmarking tools, as an external audit against the Guideline's standards will immediately indicate a benchmark against international standards, or national standards in the case of ANSI Z10-2012.

Official, formal audits are conducted by third party external organizations, who have auditors trained to conduct formal audits. This formal measurement also acts as a form of recognition for safety efforts. A level of performance is awarded after the audit and in some instances, accreditation is given.

SELF-AUDIT

A self-audit is when internal accredited safety auditors, in conjunction with management, evaluate all elements of the safety system against the standards, quantify the results, and produce a measure, normally a percentage, of compliance. The self-audit should be conducted on a six-monthly basis and all levels of management should be involved in the auditing process. The self-audit is also a self-truth session and allows for a close examination of the safety system.

INFORMAL AUDIT

An informal audit is conducted on specific system elements and provides an interim status report as to the work being done to control loss. A limited audit only involves a few specific safety program elements such as permits, written safe work procedures, inspections, etc. This report measures these elements against the predetermined standards, quantifies them, and gives direction for action plans to be compiled.

FORMAL AUDIT

A formal audit follows a structured evaluation system, a comprehensive quantification, and culminates in a detailed written report highlighting strengths and weaknesses of the safety system. The report becomes a working document, and authority, responsibility, and accountability to modify and rectify weaknesses found can be delegated. Time limits for action are recommended by this report.

THE AUDIT PROTOCOL

The audit protocol is a guide for safety system auditors which helps them audit against a specific standard. The audit protocol could be drawn up to cover the Guideline used. This means it summarizes the requirements of the Guideline into measurable and auditable elements, identifies documents which need to be reviewed, processes to be examined, and allocates a score to each element for scoring purposes.

Protocols can be written to cover all aspects of a company's safety management system so that when audited, the system is audited against its own standards. Legal compliance audit protocols compiled from local safety regulations can also be compiled and used.

An audit protocol document should contain the following:

• Name of standard
• Minimum requirements
• Minimum standard details
• What to look for during the inspection to measure conformance
• Instructions as to how to test the system
• What evidence is required for the element detail
• Questions to ask during the verification process
• The scoring scale for the minimum standard detail

MEASUREMENT

The audit protocol should allow for the auditors to allocate a measurement or score against each element, so that the compliance and achievement of the standards for that element can be quantified. Many audit protocols use a "Go" or "No Go" score. This means that if there is compliance, a number is allocated, and if there is a deviation, no points are given. Some rank fulfillment of the requirements on a 1–10 scale, where (10) indicates complete compliance and (0), non-compliance to the requirements of the standard. Others again use a 1–5 scale of measurement. A number must be allocated as managers understand numbers of measurement and what gets measured gets attention (Figure 6.2).

ELEMENT	PTS	QUESTIONS THAT COULD BE ASKED	VERIFICATION	WHAT TO LOOK FOR?
2.8 RISK ASSESSMENT (HIRA)				
Hazard identification done?	5	What method of hazard identification was used?	Copies of inspections reports/checklists used/incident recall/observation reporting/formal HIRA system. Copies of risk assessment forms/matrixes, etc.	Note possible hazard situations during the inspection. Is their system identifying hazards sufficiently?
Risk Assessments completed?	5	How many formal risk assessments were conducted during the last 12 months?		Site, process, new equipment, new process, general, risk assessments
Probability, Severity and Frequency considered?	5	Was a matrix used. What numbers were used. What were the highest risks identified?	Copies of ProbabilityX SeverityX Frequency and end numbers. There must be a ranking system.	Evidence that Risks are been identified, ranked and acted on.
Risk register kept as per Company Standard?	5	Is an updated risk register being kept?	See a copy of the Risk Register as per Element 4.2	Evidence of Risk Controls being put in place.
Action plans and follow-up compiled?	5			
Re-assessment of the risks identified?	5	Follow-up actions on the risks identified. Has something been done about it?	Proof that action has been taken to reduce risks identified. Maintenance order/pictures/before and after. New risk score compared against original assessment score.	During the inspection look to see if the risks you identify are on their register.
Total	0 30			

FIGURE 6.2   An example of an audit protocol.

The audit protocol must be appropriate for the organization being audited. If the audit is to measure against a Guideline's standards, then it must be aligned to these standards and should not deviate from them. If a protocol is used for organization XYZ, then it should be drafted to cover all of XYZ's safety standards. Bear in mind that applicable local safety legislation requirements would be the minimum to strive for and would be included in the audit protocol.

ELEMENT RISK OR BENEFIT WEIGHTING

Each element should have a risk weighting, or benefit indicator, which is used for the scoring. The weighting could be on a 1–10 scale with the high-risk/benefit elements being ranked a (10) and other elements similarly weighted according to the risk they pose, or their contribution to risk reduction. The weighting is used to multiply the final score of the element, so that different elements are scored according to conformance and risk or benefit weighting (Figure 6.3).

As per the example, Machine Guarding's risk/benefit would rank (10); Safety Talks (3); Work Permits (9); Safety Notice Boards (2), and so on. If the protocol scores all *minimum standard details* on a 1–5 scale, the maximum points for machine guarding will be the number of *minimum standard details* multiplied by five points each. If there are eight *minimum standard details*, the maximum point for the element is $8 \times 5 = 40$. The maximum score of 40 is then multiplied by the weighting of 10, which gives the maximum possible score for that element as 400 $(40 \times 10 = 400)$. Safety talks have a maximum score of 25 because the element has 5 *minimum standard details* each carrying 5 points. The weighting is 3, therefore the maximum total points for the element safety talks is (25×3) or (score × weighting), giving the maximum possible points of 75.

THE SAFETY SYSTEM AUDIT PROCESS

Appointed auditors conduct a thorough physical inspection and examination of the entire work area. Relevant control documents are scrutinized, activities are monitored, and safety systems tested. The auditors critically examine each facet of the

Element	Number of minimum standard details	Score (1–5)	Risk/benefit weighting (1–10)	Maximum total points
Machine guarding	8	$8 \times 5 = 40$	× 10	400
Notice boards	4	$4 \times 5 = 20$	× 2	40
Safety talks	5	$5 \times 5 = 25$	× 3	75
Injury statistics	7	$7 \times 5 = 35$	× 2	70
Permits	6	$6 \times 5 = 30$	× 9	270

FIGURE 6.3   An example of risk or benefit weightings for 5 Elements.

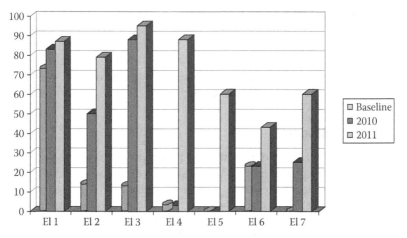

Element 1	Injury/occupational disease records	Element 4	Internal event reporting and investigation
Element 2	Internal accident reporting and investigation	Element 5	Damage (non-injury) statistics kept
Element 3	Injury/occupational disease statistics	Element 6	Insurance: apportioning of costs
		Element 7	Near miss incident and accident recall

FIGURE 6.4   Audit scores for Elements 1–7 over a 3-year period.

safety system, test the system, interview employees, and quantify the work being done to eliminate hazards and reduce risk (Figure 6.4).

TESTING THE SYSTEM

Part of the audit is the testing of the system. Critical activities and areas are inspected and scrutinized. Requirements of standards are checked against what is happening in the field. For example, if the standards call for a lock and a tag to be placed on an isolator, circuit breaker, or a disconnect of a circuit being worked on, then this is tested by visiting the switch room and observing if locks and tags are in place as required.

EMPLOYEE INTERVIEWS

Part of the audit process is arranging formal interviews with employees to ascertain their involvement in, and how the safety system affects them. This interview must be conducted in a friendly, professional manner and must not be used as a fault finding exercise. Questions must be structured and specific. The same questions should be asked to all employees interviewed, so that the auditors can form a picture of the employees' viewpoint of the safety system. Employees being interviewed should be asked for any suggestions to improve safety at the workplace and should be thanked for their input.

QUESTIONING TECHNIQUE

When auditing the correct questioning technique must be used to gather as much information about the system in as short a time as possible. Incorrect questioning technique can lead to a waste of time and incorrect information being received. There are four main types of questions:

- Closed questions
- Open questions
- Probe questions
- Mirror questions

A combination of all types of questions could be used to derive information needed to quantify the safety system.

- *Closed questions* are very direct and to the point. They restrict the possible responses. The normal reply to a closed question is either yes or no.
- *Open questions* establish a broad topic area and allow the respondent to participate freely and are very useful in all types of interviews. Sometimes, open questions can be very time consuming as the respondents may discuss irrelevant topics.
- *Probe questions* are used to gather more information concerning the element. They normally ask the respondent to clarify and enlarge on what they have just said. When using probing questions, one must be aware of the sensitivity of the respondent's feelings.
- *Mirror questions* restate the respondent's last comment for clarity. They give the respondent a chance to hear again what they have just said and gives them the opportunity to check that they meant what they said. Mirrors are very effective in avoiding misunderstanding and gaining clarity. When using various types of questions, it should be remembered that effective listening is as important as effective questioning.

WHO SHOULD CONDUCT AUDITS?

Internal audits should be conducted by personnel who have been trained as safety system auditors. As many employees as possible should be involved in the safety auditing process. This provides further training. A more objective audit is obtained by using different auditors for each audit. Safety and Health Representatives should be used extensively for internal audits, and they can also participate in the annual audit of the entire system. Safety and health consultants can be used to conduct preliminary or baseline audits. Safety practitioners from similar organizations could also be invited to participate in the audit process. Management and employees should be well represented during audits and the Safety Coordinators and Risk Managers should be involved.

THE AUDIT PROGRAM

When embarking on a formal audit, the audit program will follow a certain procedure (Figure 6.5). The order could be as follows:

PRE-AUDIT DOCUMENTATION

The submission of pre-audit documentation to the audit team sometime before the
audit is very helpful in that it can enlighten them as to the following:

- Number of employees
- Number of contractors employed
- Injury rates
- Production flow lines
- Energy used (water, electricity, gas, nuclear, etc.)
- Organization organogram
- Safety policy
- Safety department structure
- Risk assessments
- Size of work and outlying areas, etc.

The pre-audit document saves a lot of time during the opening conference, as
the auditors are already aware of many aspects of the organization and its safety

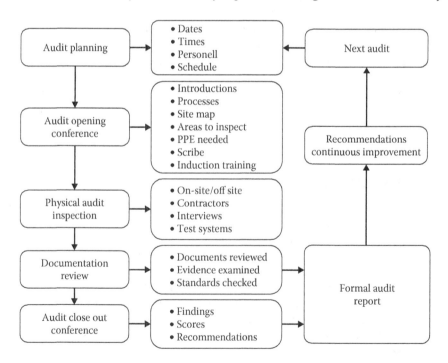

FIGURE 6.5 The safety system audit process.

management system. The auditors can also plan the inspection better based on information gathered from the pre-audit information.

AUDIT OPENING CONFERENCE

An audit opening meeting is where introductions take place, the audit plans are agreed to, where various element coordinators are identified, physical areas chosen for the inspection, and selected employees to be interviewed are nominated. This opening conference will also act as a forum to ascertain who will be accompanying the auditors, who will take notes during the inspection, the types of personal protective equipment required, and any other formalities that need be discussed.

During the pre-audit meeting, the tour through the facility will be planned. The first day of the audit may also entail examining the accident records, injury statistics, accident, and near miss incident investigation and recording systems, so that high-risk areas or processes can be identified. The safety system of the organization should be explained to the auditors, and in some instances, a brief presentation could be held to inform the auditors about the organization and the safety system.

PHYSICAL INSPECTION

The next step in the auditing process is the physical inspection. The auditors may choose to spend between 2 and 4 days carrying out physical inspections to ensure that the entire work site, and selected outlying areas, are covered. Time taken will depend on the size and nature of the facility being audited.

Random Sampling

In some instances, it may not be feasible, practicable, or possible to inspect every square inch of the work areas. Some outlying facilities may be far from the main works and travel time will absorb most of the audit time. In these cases, random sampling techniques are used. During the audit opening conference, the auditors select certain outlying areas to be included in the inspection. What is found in these random selected areas is used as a representation of the other areas.

In other instances, for example, not all flammable liquid stores can be inspected, so a few are selected for inspection as a random sample. The same technique applies to the inspection of change rooms, kitchens, electrical rooms, and other areas. Inspecting each and every one would take too much time and not be feasible. Once selected for random sampling, both the organization and the auditors accept that those areas selected will be a cross cut representation of the entire organization.

How to Do an Audit Inspection

Most inspectors make the mistake of walking through a work area too fast. It is beneficial to simply stand at the entrance to the workplace and observe the people, machinery, environment, material, and work process. Note what work the employees are doing in the area, the protective equipment being worn, and the environmental factors such as the lighting, ventilation, noise, and temperatures.

During the inspection, ask the questions, "Is this safe, can this cause an accident?" And, "Is this according to your safety management system standards?" Auditors should try to cover the entire work area and note and record deviations. All hazards must be noted and brought to the attention of the employees in the area being inspected. While doing the inspection, auditors should be curious. Look inside cupboards, under workbenches, on top of mezzanine floors, and look behind articles.

Auditors should note the area in which the deviation was found or where compliance to standards was found. They should maintain a positive attitude during the inspection. Smile, be friendly, and compliment employees wherever possible. An audit is a learning experience for both the auditors and the organization. Auditors record the deviations and where they occurred, and if necessary, ask the organization's representative to react immediately and have "A" class hazards rectified as soon as possible.

The audit team should keep an open mind and not be biased during the inspection. It is important that all deviations are classified under the various elements, so that the evaluation can be conducted easily. The key to a successful inspection is, "write it down."

It is important to set an example by wearing the corrective personal protective equipment during the inspection and not to commit high-risk acts such as failing to observe a warning or mandatory sign.

DOCUMENTED EVIDENCE REVIEW

A major part of the audit procedure is the documentation review or verification session. This is where documented evidence in the form of completed checklists, copies of permits, procedures, and standards are inspected to verify that the organization is following its safety system standards and that they have documented evidence of this.

Evidence Preparation

In preparation for the audit, the safety system should retain copies of evidence so that they can be presented to the auditors during the documentation review sessions. The documentation should cover the preceding twelve months' activities. Future improvements to the safety system should not be presented as only the activities that were carried out during the past 12 months can be evaluated and quantified.

Evidence can be in the form of hardcopy or electronic records. During the evidence review sessions, many organizations have computers available and the document or evidence under review is projected onto a screen for the auditors' scrutiny. Electronically signed documents and checklists are as acceptable during audits as hardcopies are. Each element should have a reference number for tracking purposes.

IS THE SYSTEM WORKING?

As with the site inspections, random samples of documented evidence only need to be reviewed. Evidence based on the inspection is examined to ensure that the system is working. During the inspection, notes are taken on items that have been inspected or work procedures observed. During the review session, the completed checklist for the item seen during the inspection is examined to see if the item was in fact inspected as per standard.

If procedures are noted during the inspection, the written procedure and proof of employee training on the procedure would be requested and scrutinized. The important question to be ascertained during the review of evidence is, "Is the system working?"

AUDIT CLOSEOUT CONFERENCE

Once all the documentation has been reviewed and systems are verified, a feedback session is held with all concerned parties. Here, the auditors present feedback on the strengths and weaknesses of the system, and at the conclusion, announce the score for each section and the total for the entire safety system.

AUDIT REPORT

A few weeks after the audit, a formal audit report is tabled and discussed by the auditors. This feedback session will include the drawing up of an action plan to rectify the weaknesses found during the audit. Management will be presented with a prioritized action plan listing what needs to be done. This is the "Act" stage of the *Plan*, *Do*, *Check*, *Act* methodology.

SUMMARY

Many safety systems measure consequence, which is post-contact control and reactive safety. Safety management system audits evaluate and measure every aspect of work being done to reduce risk and prevent accidental loss, and are ideal means of measuring precontact control efforts. Management is more inclined to pay attention to anything that is quantified and the audit is ideal as it puts a measure to the degree of conformity to standards. Only by conducting audits on a frequent basis can an organization incorporate a process of continual safety system improvement.

7 Safety Leadership and Organization—Part 1

SAFETY SYSTEM COMPONENTS OR ELEMENTS

A safety management system (SMS) consists of a number of items, systems, sub-systems, processes, and activities, prescribed by elements, which need to concur on a regular basis in order to provide an ongoing system of risk control. The reduction of risk is the only way to eliminate losses caused by accidents.

Many of these elements (i.e., processes or programs) are required by safety and health legislation and are not new or unknown activities. They must, however, be integrated into the day-to-day management of the organization to be effective and sustainable. These elements will vary from workplace to workplace even though there are some core elements that are common and applicable to most organizations. A comprehensive safety management system normally comprises sections, elements, minimum standards, and minimum standards detail, and could consist of at least 70–80 elements. To explain the sections and elements of a safety management system, an example safety management system (Example SMS), is used.

EXAMPLE SAFETY MANAGEMENT SYSTEM (EXAMPLE SMS)

The Example SMS is based on the original National Occupational Safety Association (NOSA) 5-Star Safety Management System, which has been extensively modified to encompass most of the requirements of most of the Guidelines (NOSA, 1995). The Example SMS consists of 5 sections and a total of 84 elements. Section 1, *Safety Leadership and Organization* of the Example SMS, consists of 30 elements. The first 5 elements will be discussed in this chapter and the last 25 elements will be discussed in Chapter 8 (Figure 7.1).

SECTIONS, ELEMENTS, AND SUB-ELEMENTS

All elements have prescribed *minimum standards* as well as *minimum standard detail*. Many elements have sub-sections. The five sections of the Example SMS are: *Safety Leadership and Organization* (Section 1), *Electrical, Mechanical and Personal Safeguarding* (Section 2), *Emergency Preparedness and Fire Prevention* and *Protection* (Section 3), *Accident Reporting and Investigation* (Section 4), *Physical Workplace Environment* (Section 5).

One of the elements of the *Safety Leadership and Organization* section is that of Safety Committees. One of the *minimum standards* of this element would be

Number	Element title	Number	Element title
1.1	Managers Responsible for Safety and Health	1.16	Safety Newsletters
1.2	Safety Policy: Management Involvement	1.17	Safety and Health Representatives
1.3	Safety Performance Indicators (SPI)	1.18	Safety Management System Audits
1.4	Safety Committees	1.19	External Third Party Audits
1.5	Management of Change	1.20	Safety Publicity Boards
1.6	Safety and Health Training	1.21	Publicity, Bulletins, Newsletters, etc.
1.7	Work Permits	1.22	Safety Competitions
1.8	Organization Risk Management	1.23	Toolbox Talks, Safety Briefings, etc.
1.9	Written Safe Work Procedures	1.24	Safety Specifications:
1.10	Planned Job Observation	1.25	Safety Rule Book
1.11	Safety Inspections	1.26	Safety Reference Library
1.12	Safety Suggestion Schemes	1.27	Public Safety
1.13	Employee Job Specifications	1.28	Annual Report – Safety and Health
1.14	Medical Examinations	1.29	System Documentation Control
1.15	Off-the-job Safety	1.30	Continual Improvement

FIGURE 7.1   The 30 Elements of Section 1, of the Example SMS.

the formation of committees at various levels or departments within the organization. Another *minimum standard* would be different types of committees needed, for example, permit committees, accident investigation committees, etc.

A *minimum standard detail* of the first minimum standard would be the keeping of minutes of meetings. Another *minimum standard detail* would be, for example, that the minutes are signed by the chairperson (manager or supervisor) of the committee.

Each minimum standard detail has a score allocated for auditing purposes. The Example SMS ranks each minimum standard detail on a 1–5 scale, and the main element has a weighting of 1–10 scale, dependent on the risk ranking of the element or the benefit derived from the element.

Example SMS

Section: *Management Leadership and Organization* (Section 1).
Element: *Safety Committees.* (Section 1, Element 1.4) (S1, E 1.4) (Weighting 4).
Minimum Standard: *Committees formed to meet the legal requirements, or organization's need.* (20).

- *Minimum Standard detail 1*: All departments to have a safety committee. (1–5)
- *Minimum Standard detail 2*: Committees to meet monthly. (1–5)
- *Minimum Standard detail 3*: Minutes to be kept of each meeting. (1–5)
- *Minimum Standard detail 4*: Minutes to be signed by chairperson of the committee. (1–5)

Total maximum score of the element (*Safety Committees*) is: 20 points (4 minimum standard detail with a maximum of 5 points each). Weighting 4, for example, possible maximum score: 80 points (20 × 4).

ELEMENT STANDARD REQUIREMENTS

Each element of the safety system should have a written standard which can be stored in hardcopy or as an electronic file. Each standard should have: a title, an element number, a date, revision number, legal reference, sources and references, signature of authorizer, name(s) of compiler(s), and reviewers' signatures. The standard should also contain an objective, minimum standard requirements, minimum standard details, the process, measurable criteria, and it must allocate responsibility for actions.

CORE (COMMON) ELEMENTS OR COMPONENTS

Core elements form the basic safety management system elements. These are elements common to most workplaces, industries and mines, and include elements such as the following:

- Work permits
- Safety committees
- Management responsibility
- Employee participation
- Business order (Housekeeping)
- Electrical safety
- Motorized transport safety
- Occupational hygiene
- Accident investigation
- Personal and mechanical guarding
- Fire precautions
- Inspections and many more

PRINCIPLE OF THE CRITICAL FEW

The principle of the critical few states that *a small number of basic causes could give rise to the majority of safety problems.* Critical elements could be responsible for the majority of injury causing accidents. These critical few items (critical safety elements) should receive maximum safety control to minimize their potential for causing (the majority) of problems. Precontact control means directing safety efforts towards controlling these crucial areas before a loss occurs.

What Are Critical Elements?

Critical safety elements are elements, identified by the hazard identification and risk assessment process that pose the highest risk to the organization. These may include environmental and employee factor elements, which need to be controlled constantly to prevent losses occurring.

Critical safety elements are those elements most likely to give rise to losses. Past experience based on thousands of safety inspections and audits have shown that control over certain aspects of the work environment and work practices can significantly reduce accidents. Critical elements would have a higher weighting allocated to them for auditing purposes as a result of their risk ranking.

Examples

Critical safety elements are elements that reduce risks posed to employees and which need to be managed on an ongoing basis to prevent accidental loss. Critical elements may differ from workplace to workplace and could include elements such as the following:

- Hazardous material handling
- Confined space entry
- Work at heights
- Machine guarding
- Ergonomics
- Radiation safety
- Rescue procedures
- Underground ventilation
- Fire prevention, etc.

Risk assessments will identify which safety system elements are critical to the organization and which are not so critical but also important.

Why These Elements?

Experience has shown that there are between 70 and 80 critical safety activities (elements) that need to be in operation to constitute an effective safety management system. Most of these are processes common to most workplaces. Dependent on the nature of the business and its operations, these common elements will form the basis of the safety management system. These elements may vary from organization to organization and from industry to industry. The emphasis on individual elements will also vary according to the nature of the process, culture of the workforce, and category of business such as mining, the iron and steel industry, transportation, the fishing industry, manufacturing, construction, etc.

Benefit

The benefit of controlling critical safety elements, is that the work being done to manage safety is channeled at reducing the risk and potential loss in areas that have been identified as crucial. Some critical safety elements help control the health and safety of the work environment, which would contribute to the reduction of losses due to an unsafe workplace.

Other critical safety elements are directed towards the protection of the employees within a workplace. These controls would include items such as critical task procedures, rules, training and activities that involve, motivating, guiding, and training employees in safe work practices.

Elements that have a direct influence on the work environment include items such as the following:

- Business order (Housekeeping)
- Electrical safety
- Stacking and storage
- Ventilation
- Lifting gear
- Demarcation
- Machine guarding
- Hazardous substance control
- Lighting
- Hazardous environment controls, etc.

Elements that help to ensure safe work procedures and habits could include:

- Critical task procedures
- Rules and regulations
- Job safe procedures (JSPs)
- Appointment of Health and Safety Representative
- Safety and health training
- Regular safety meetings
- Safety communication
- Safety promotion
- Medical examinations, etc.

PRECONTACT, CONTACT, AND POST-CONTACT CONTROL

There are three areas of safety control. They are the precontact stage, the contact stage, and the post-contact stage. Precontact activities are the interventions and processes that take place before any accident occurs. They are ongoing hazard identification and risk reduction processes that occur according to the safety plan, objectives, and standards.

Most safety efforts are aimed at the post-contact stage, when a loss has already occurred. The accident has happened and there is a sudden response to now rectify the accident causing problems, that is reactive safety and not proactive safety. Control before the accident and subsequent loss, is taking action at the precontact stage and is proactive safety. Precontact management control of risk is facilitated by a structured safety management system and its components (Figure 7.2).

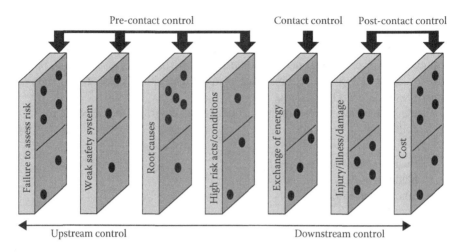

FIGURE 7.2   Precontact, contact, and post-contact control.

SAFETY MANAGEMENT SYSTEM BASIC ELEMENTS

A safety management system embraces ongoing activities and efforts directed to control accidental losses by monitoring critical safety elements on an ongoing basis. The monitoring includes the promotion, auditing and improvement of the critical and other elements constantly.

The success of many safety management systems is that they begin the implementation of the system with the tangible, visible safety elements, and then move onto the intangible elements. It is recommended that the basics of safety be put in place first and then the more intangible, long-term benefit elements be phased in next. Management and employee buy-in is easier when short-term gains are realized and they can see changes to the workplace being brought about by the safety management system. This also helps to reduce resistance to change.

BUSINESS ORDER AND LEGAL REQUIREMENTS

When faced with the task of implementing a safety system, the organization should implement processes that are required by safety legislation, and elements that will give the best return with the lowest cost, and within the shortest time. Business order (good housekeeping) is a basic requirement of international safety legislation and is a good place to start. Once the workplace, including offices, storerooms, kitchens, and all work areas are in order (a place for everything and everything in its place, always), the impact of the safety system becomes visible. Business order is the foundation of a safety management system and as one safety expert put it, "If you can't manage the workplace, how can you manage the people in that workplace?"

Getting the organization in order is a management responsibility and once it is achieved a number of hazards would have been eliminated, and the foundation established on which to build the safety management system.

MANAGEMENT AUTHORITY, RESPONSIBILITY, AND ACCOUNTABILITY

DEFINITIONS

- *Safety authority* has already been defined as the right or power assigned to an executive or a manager in order to achieve certain organizational safety goals.
- *Safety responsibility* is a duty, obligation, or liability toward safety for which someone is accountable.
- *Safety accountability* is being liable to be called on to render an account on safety and be answerable for safety issues.

MANAGERS DESIGNATED AS RESPONSIBLE FOR SAFETY AND HEALTH (SECTION 1, ELEMENT 1.1) (S1, E1.1)

MANAGEMENT SAFETY APPOINTMENTS

Many of the Guidelines recommend that senior management appoints one of its top managers responsible to coordinate the safety management system implementation and maintenance. This person is sometimes referred to as the Safety Champion. This appointee must report to the executive committee on a regular basis as to the status of the safety system and these reports must form the basis of future system improvements. This appointment should be in writing and a brief description of duties and functions should be spelt out. This appointment must be made known to the entire organization.

MANAGERS

All managers in the management line must be appointed in writing, responsible for the implementation and maintenance of the safety management system in their areas of responsibility. These appointments must spell out their safety authority, responsibility, and accountability, and must be acknowledged by the appointees.

SUPERVISORS

All supervisors, team leaders, or foremen in the management line must be appointed in writing, responsible for the implementation and maintenance of the safety management system in their areas of responsibility. These appointments must spell out their safety authority, responsibility, and accountability and must be acknowledged and accepted by the appointees.

SAFETY AND HEALTH REPRESENTATIVES

The selection, nomination, and appointment of Safety and Health Representatives (Safety Representatives) is perhaps the next most important appointment that senior management can make. In some instances, these appointments are a legal requirement prescribed by safety and labor law, and must be done accordingly.

Selected individuals from within the workplace are selected or nominated, either by management or the workforce, or both, to act as safety representatives for the forthcoming period, normally a year. They are then appointed in writing and their duties and functions are detailed in the appointment document.

The Safety Representatives should undergo suitable training to prepare them for their duties which, briefly summarized, is to conduct a monthly inspection of their work areas, identify high-risk acts and unsafe conditions, and complete the inspection report. Once the inspection is completed, they present the report to their supervisor or manager who undertakes to take action on the deviations noted.

SAFETY DEPARTMENT

The safety department appointments are done by management. Getting the right person in the right job is vital if the safety department are to be effective coordinators of the safety system. The safety department cannot, and should not, be held accountable for the safety performance of an organization. All employees have some safety responsibility, but management, senior and line, have ultimate safety authority therefore is ultimately accountable.

Safety Department Job Purpose

The main objective of the safety department should be to guide, educate, train, and motivate all levels of management, unions, contractors, and the workforce in the techniques of accident and industrial illness and disease prevention, in an ongoing effort to reduce risk to an acceptable level in order to prevent injury and illness to employees and damage to property. They also advise on, and coordinate the safety management system. The safety department should be a staff function and not a line function. To summarize their function: Guide, Educate, Train, Motivate, Advise, and Coordinate (GETMAC).

APPOINTING PROFESSIONALS

When appointing safety and health department employees, management has a duty to the profession to ensure suitably qualified and experienced practitioners are selected for the safety department. Proper job descriptions based on the American National Standards Institute (ANSI) guidelines: ANSI/ASSE Z590.2 – 2003, *Criteria for Establishing the Scope and Functions of the Professional Safety Positions*, or similar, should be used as selection and training criteria for safety staff.

Safety departments should be very professional. Their true place is advising management and coordinating the activities of an ongoing safety management system. They cannot improve the safety by accepting the responsibility for safety. They should not directly try to influence behaviors of employees. Only management can do that.

SAFETY COORDINATOR

The Safety Coordinator's (Safety Advisor, Safety Practitioner, Loss Control Professional) function should also be to guide, educate, train, and motivate all levels of management, contractors, and the workforce in the techniques of risk reduction, accident prevention, and the safety management system. They also advise on all aspects of safety and coordinate safety activities in their area(s) of responsibility. Further, they coordinate the implementation of the safety management system and offer support to management, contractors, and workers concerning the safety management system. Board certification is recommended.

INDUSTRIAL (OCCUPATIONAL) HYGIENE

Industrial (Occupational) Hygiene is the science and art devoted to the anticipation, recognition, identification, evaluation, and control of environmental stresses arising out of a workplace which may cause illness, impaired wellbeing, discomfort, and inefficiency of employees or members of the surrounding community. It can be described as the science dealing with the influence of the work environment on the health of employees and is an important part of the safety system.

Industrial (Occupational) Hygienist

The Industrial Hygienist (IH) coordinates the industrial hygiene elements of the safety management system as they call for specialist skills and training. Ideally the appointee should be a member of a recognized institute and have appropriate board certification.

Objectives

The objective of the Industrial Hygienist is to recognize occupational health hazards, evaluate the severity of these hazards, and eliminate them by recommending the instituting of control measures. Where the occupational health hazards cannot be eliminated entirely, occupational hygiene control methods aim to reduce the exposure to the hazard and institute measures to reduce the hazard.

SAFETY COMMITTEE CHAIRPERSON

The chairperson of a committee is the person who leads a safety and health committee and should be a member of management. The safety committee chairperson must also be appointed by senior management, in writing, and duties should be briefly described in the appointment document. Depending on the structure of the safety committee, or tiers of safety committees, it is ideal to have the chief executive officer (CEO) of the organization head up the executive safety committee (EXCO).

ACCIDENT INVESTIGATORS

Accident investigators should be appointed in writing and must receive the necessary training in accident investigation techniques. The investigator of an accident, or high potential near miss incident, should ideally be the supervisor of the area in which the event took place. Since safety is a line management function, line management should take ownership of accidents in their areas and investigate them. The safety department can assist in the investigation, or in major events, they may be called on to take the lead. Special accident investigation committees should also be appointed where and when necessary.

FIRE OFFICIALS

The fire department members or the fire coordinator must also be trained and appointed to coordinate the elements of the safety system dealing with fire protection and prevention. Fire wardens, fire marshals, equipment inspectors, as well as emergency coordinators must also be appointed as per the standards of the safety management system elements.

OTHER APPOINTMENTS

Other appointments could include issuers and receivers of work permits and others, once again, dependent on the needs of the organization. Management must ensure that all appointments are done in writing so that safety authority, responsibility, and accountability are allocated to the correct people. Their duties must also be spelt out in their letters of appointment.

OCCUPATIONAL SAFETY AND HEALTH POLICY (THE POLICY STATEMENT) (S1, E1.2)

The Chief Executive Officer (CEO) must approve the contents of the Safety and Health Policy Statement (*the* safety policy) and ensure that it is updated in accordance with the needs of the organization. Since the policy is the initiating document of the safety system process, it should

- Include a commitment to injury and ill-health prevention.
- Indicate safety responsibilities and accountabilities.
- Refer to continuing improvement of safety and health initiatives.
- Contain a commitment to comply with safety and health legislation.
- Provide a framework for safety objective setting.
- Be documented, displayed and maintained.
- Be freely available.
- Be communicated to all affected parties.
- Receive periodically review.

This is the declaration that drives the safety management system, and is viewed as the most important document in the system. The policy statement should be made known to all employees, contractors, visitors, and other interested parties. The policy must be signed by the CEO. In some organizations the entire executive management sign the policy.

The policy should be framed and displayed throughout all workplaces, including offices and foyers and meeting rooms. A copy should be reproduced in the employee safety handbook and a copy included in contract bid documents. Since it forms the framework for setting safety objectives, it should be reviewed at least annually.

SAFETY PERFORMANCE INDICATORS (SPIs)
AND OBJECTIVES (S1, E1.3)

Safety objectives should now be established for the organization. The status of the safety within an organization must constantly be measured against the company's safety goals and objectives. Safety performance indicators (SPIs) should be upstream indicators that can be managed, achieved, and measured. Collectively they form part of the safety system action plan.

MEASUREMENTS

Safety management systems should have measurements based mainly on precontact objectives which are manageable and achievable. Safety measures of precontact achievements are more meaningful, and organizations should not only focus on the post-contact or lagging indicators. The Guidelines suggest that safety measurements be both proactive and reactive, both upstream and downstream, so that a clear picture of the safety efforts is obtained.

Upstream measurements could include the following:

- Number of near miss incidents reported
- Safety committee meetings and attendance
- Toolbox talks held
- Evacuation drills held
- Internal audit scores
- Safety observations rectified
- Safety inspections done
- Safety Representative inspections done and checklists submitted
- Housekeeping scores, etc.

Downstream measurements could include the following:

- Fatality rate
- Injury rate
- Injury severity rate
- Property damage costs
- Fire losses
- Business interruptions, etc.

PROACTIVE, UPSTREAM SAFETY PERFORMANCE INDICATORS

Manageable safety performance indicators should primarily be based on proactive, upstream objectives. An organization should break from traditional reactive measurements and shift the focus to upstream, precontact, and doable actions that contribute to the ongoing reduction of risk. These proactive safety indicators can also be cascaded down to all levels of management and can be measured on a short, medium and long term basis. Achievement of these SPIs creates safety accountability at all levels.

Although they should be considered, downstream measures, over the years have been subject to manipulation and misinterpretation to improve so-called safety performance and are not reliable measures of safety efforts. They are difficult to set as manageable targets, as the obvious target is "zero injuries and fatalities."

SAFETY AND HEALTH COMMITTEES (S1, E1.4)

A safety and health committee (safety committee) can be defined as a group that aids and advises both management and employees on matters of safety and health pertaining to company operations. Furthermore, it performs essential monitoring, educational, investigative, and evaluation tasks.

A safety committee is one way to ensure the active participation of a large number of employees and also to ensure that supervisors, foremen, employees, and management are involved and informed about the safety management system. Employees and managers become involved in the safety process as participation in the safety committee activities allows for opinions to be heard, and suggestions to improve safety can be discussed. Committees should operate in all departments throughout the organization, so that participation and communication reach all levels within the company.

SUPPORT

Safety committees must have the full support and backing of top management otherwise they are doomed to fail. Management must be sincere about its support of the safety system, and show this support by holding regular executive safety committee meetings and by taking action recommended by sub, or special committees.

TYPES OF COMMITTEES

There are numerous types of safety committees. These are: joint committees, departmental committees, maintenance committees, committees consisting of Safety and Health Representatives, the Executive Safety Committee (EXCO), safety suggestion committee, and other ad hoc committees which may be constituted from time to time. These may include committees formed to help organize the safety day, or organize the annual housekeeping competition, or prepare certain elements of the safety system that need modification, rectification, or implementation.

Executive Safety Committee (EXCO)

The executive level safety committee chaired by the CEO of the organization, comprising top management, should meet monthly. This is the policy-making body and the group that set standards and direction for the safety management system. The executive appointed for safety coordination reports his or her findings, suggestions, and recommendations concerning the safety system to this committee. Without guidance and direction from such an EXCO committee, the safety system may not be successfully implemented or maintained.

Departmental Safety Committees

The heads of departments should be members of the executive committee so that all departments of the organization are represented. These departmental heads in turn should chair their own departmental safety committees, which means that all departmental employees are represented by the safety committee system. Having the departmental head or supervisor chair the meetings means that certain decisions can be taken at the committee level, directions given, and two-way communication between workers and management takes place.

Numerous safety committees have failed because the committees have been chaired by the wrong level of employees, without the correct authority. Safety committees must be chaired by a manager, foreman, team leader, group leader, or head of that particular department, as only they have the authority to take decisions and bring about change. Safety department personnel should not chair safety committees but should act as facilitators and secretaries to the committees.

Special Committees

Special committees can be formed to operate on the same lines as the structured safety committees, and these committees could be temporary until the project is completed, or the specific objective has been achieved. Types of special committees could include the following:

- Accident investigation committee
- Safety competition organizing committee
- Safety education committee
- First aid training committee
- Personal protective equipment selection committee
- Ergonomic project review committee
- Written safe work procedure drafting committee
- Standards review committee
- Pre-audit team committee
- Housekeeping competition adjudication committee, etc.

Joint Safety Committee

The joint safety and health committee is advocated by safety legislation in a number of countries. The joint committee concept stresses cooperation and a commitment to safety, as a shared responsibility, both by management and employees. At joint committee meetings, employees can become involved in safety discussions and their ideas can be translated into actions. The committee serves as a forum for discussing regulatory changes, safety system elements, processes, or new and unusual procedures that could give rise to accidents.

The concept of joint decision making in safety is facilitated by joint safety committees as management and workers can now face each other across the same meeting table with a common ground and a common agenda based on the prevention of loss causing events. Representation by management at a joint committee clearly indicates commitment to safety and visible involvement.

Accident and Near Miss Incident Investigation Committee

In many instances, a special accident investigation committee is constituted to investigate certain major accidents. The main function of the committee is to investigate the accident and report its findings and recommendations to top management. In some countries, this committee is required by legislation.

Safety and Health Representative Committees

A committee comprising of Safety and Health Representatives and other interested parties is sometimes required by local safety legislation. This is also a vital safety committee which can contribute greatly to the safety system, as it creates a platform for open communication between management and employees concerning employee safety and health.

FUNCTIONS OF SAFETY AND HEALTH COMMITTEES

The main functions of safety committees are:

- Meet regularly (preferable monthly).
- Ensure that safety momentum is maintained.
- Provide for two-way safety communication.
- Solve certain occupational health and/or safety problems.
- Propose improvements to the safety system.
- Act as a selection committee for suppliers of personal protective equipment.
- Assist in accident investigations.
- Sort and select safety suggestions and recognition schemes.
- Assist in planning safety campaigns, competitions, etc.
- Constantly monitor the safety system effort.
- Discuss near miss incidents, injury and loss statistics, and monitor trends.
- Preview new safety training programs which are about to be introduced.
- Contribute to the general improvement of the safety system.

SAFETY COMMITTEE CONSTITUTION

Most safety committees are constituted formally within the organization, and a basic constitution is drawn up governing the duties, rights, powers, and functioning of safety committees. A simple safety committee constitution could consist of the following:

- Scope of activities
- Extent of the committees' authority
- Basic objectives of the committee
- Membership
- Role of the Chairperson
- Secretarial and other functions

The procedures are also laid down in the constitution, and this would also stipulate how many members are necessary to form a quorum, the frequency of meetings, which records are to be kept, and the order of business.

COMMITTEE PURPOSE

The constitution described above should never detract the committee from its most important and basic function. The basic function of every safety committee is to create and maintain interest in safety and health, create involvement and participation, and provide a platform for two-way communication, thereby contributing to the safety management system.

RECOGNITION

In some plants, it is a practice to display the photographs of the safety committee members on the safety notice board and to also issue each member with a badge stating, "Safety Committee Member." It should also be widely publicized that any employee can speak to a safety committee member, should there be a specific problem, or should a hazard arise. This system also gives importance to being a safety committee member, and normally great competition takes place every year to be nominated or re-nominated as a safety committee member.

MANAGEMENT OF CHANGE (S1, E1.5)

Ineffective management of change has been the cause of many major accidents. Change at the workplace is inevitable and often brings progress. But it can also increase risks that, if not properly managed, may create conditions that could lead to injuries, property damage, or other losses. The main types of change within an enterprise are: changes to equipment, infrastructure, processes or product, changes in personnel, use of different material, and changes brought about by the safety and health management system.

The hazards encountered and risks posed by these changes should be assessed and the necessary controls put into place before the change, to reduce the probability of the change resulting in accidental loss.

Ideally, no modification should be made to any plant, equipment, control systems, process conditions, operating, or safety and health procedures, without authorization from a responsible manager. In some cases third party approvals may be necessary.

The management of change process could follow the following phases:

Origination: This is where the change is generated in the form of an improvement, an idea or a solution to a problem.

Review and appraisal: The changes should be reviewed and the technical, operational, safety, environmental, quality, and economical aspects of the change be evaluated. In some instances specialists may be called in for this change review.

Change approval: A document in the form of a *management of change* document, should list all the details of the change and all concerned departments and managers should approve and sign off on the change proposal and its ramifications. Technical, operational, financial, health, safety, and environment departments should be involved in the approval process.

All foreseeable hazards and risks posed by the proposed change should be considered at this stage, and safety system controls recommended. A responsible person should be appointed as the person in charge of the change project.

Implementation and verification: The change is now implemented according to a plan of action based on the approval document. The change must be verified to ensure that it is in accordance with the change document and all relevant requirements prior to restarting the changed process. All proposed safety and health controls must be in place prior to the change.

Verification documentation: All documentation relevant to the change needs to be updated. This includes: specifications, procedures, emergency actions, training manuals, checklists, etc.

Training: Affected employees and contractors need to be trained on the impact of the change prior to the restart of the changed process.

Checklists: Checklists should be used during all phases of the change to ensure the plan is being followed and necessary controls are in place and working.

Management of emergency changes: Management of change due to emergencies should also follow a sequence and receive the same attention as routine changes. These changes may be necessary in case of major events, immediate threat to employees, contractors or members of the public, or other situations that need immediate change to reduce risk.

8 Safety Leadership and Organization — Part 2

EXAMPLE SMS: SECTION 1

Section 1 of the Example SMS contains 30 elements. The next 25 elements are discussed in this chapter (Figure 8.1).

SAFETY AND HEALTH TRAINING (SECTION 1, ELEMENT 1.6) (S1, E1.6)

Lack of knowledge or skill is one of the root causes of accidents. Safety training creates an awareness of what causes accidents, what constitutes injuries and diseases, and how they can be prevented. Safety training clarifies the difference between near miss incidents and accidents, and explains safety authority, responsibility and accountability and the role that employees play in the safety management system (SMS). Safety training carries the message that the majority of accidents are caused by root and immediate causes, and are preventable. It helps teach employees how to do their job correctly and indicates what dangers may be present in their work environment.

Safety training is of vital importance to both management and the workforce as it forms an important safety knowledge base on which to build a comprehensive safety management system. Safety training workshops or classes should be held for individuals, safety committees, small groups, and work task teams on a scheduled and regular basis.

TYPES OF SAFETY TRAINING

Safety training can take many forms and can either be general, specific, or required by safety legislation, such as annual safety refresher training and training on specific aspects of the work. Specific training is required for handling of hazardous materials, lockout, issuance, and receipt of work permits, etc.

Safety Induction or Orientation

Safety induction (orientation) training is normally presented to new employees and contractors to ensure that they are familiar with the work environment and the risks inherent in the workplace and work they do. The organization's safety and health policy is also explained during these sessions. In some instances, visitors to the site

Number	Element Title	Number	Element Title
1.1	Managers Responsible for Safety and Health	1.16	Safety Newsletters
1.2	Safety Policy: Management Involvement	1.17	Safety and Health Representatives
1.3	Safety Performance Indicators (SPI)	1.18	Safety Management System Audits
1.4	Safety Committees	1.19	External Third Party Audits
1.5	Management of Change	1.20	Safety Publicity Boards
1.6	Safety and Health Training	1.21	Publicity, Bulletins, Newsletters, etc.
1.7	Work Permits	1.22	Safety Competitions
1.8	Organization Risk Management	1.23	Toolbox Talks, Safety Briefings, etc.
1.9	Written Safe Work Procedures	1.24	Safety Specifications:
1.10	Planned Job Observation	1.25	Safety Rule Book
1.11	Safety Inspections	1.26	Safety Reference Library
1.12	Safety Suggestion Schemes	1.27	Public Safety
1.13	Employee Job Specifications	1.28	Annual Report – Safety and Health
1.14	Medical Examinations	1.29	System Documentation Control
1.15	Off-the-job Safety	1.30	Continual Improvement

FIGURE 8.1   Section 1 of the Example SMS contains 30 Elements.

are subjected to a shorter version of the employee orientation, and any person spending more than a shift on site, should attend the full induction program. Multinational companies present the safety induction in different languages to cater for the multi lingual workforce.

Refresher Training

Refresher training is normally a follow up on induction training, and should be attended on an annual basis, or when an employee returns to the workplace after a leave of absence. Training in the philosophy of safety is important as it helps people understand the cause of accidents and the effects of the results. The safety system and its function forms part of most safety training classes. Training employees to understand safety standards is another vital aspect of safety training. The training program could include teaching the safety rules of the organization as well as the site's do's and don'ts. All supervisory staff as well as management should attend training in basic safety management as well.

Other Training

First aid training and training in cardiopulmonary resuscitation (CPR) and mouth-to-mouth resuscitation prepare attendees for emergencies. Rescue, evacuation training, and firefighting courses are beneficial to both the individual and the organization.

Safety training may be specific and could cover topics such as the following:

- Accident investigation
- Near miss incident recognition
- Hazardous material handling
- Work permit issuer and receiver
- Training-the-trainer
- Confined space entry
- Lifting gear safety, etc.
- Regulatory safety requirements
- Critical/Hazardous task
- First aid, and CPR, etc.

An important part of carrying out critical or hazardous tasks, is the training that is presented to enable the workers to understand the critical task procedure. Re-training after a job observation is the best method to ensure that individuals follow the critical task procedures.

Ongoing safety training for safety department staff is important and would help keep them updated on modern safety techniques such as the following:

- Risk management
- Risk evaluation
- Globally harmonized labeling system
- Ergonomics
- Sick building syndrome
- Carpal tunnel syndrome, etc.

WORK PERMITS (S1, E1.7)

Work permits are written authorizations given to employees who carry out hazardous tasks, or who work in hazardous situations (confined spaces, at heights, etc.). Work permits are documents that authorize persons to perform work in certain areas, or work on specific equipment and processes, which are inherently hazardous, have a high risk, and which require specific safety actions, both before, during, and after carrying out the work. Some permits are permission to work on site and are issued, for example, to contractors.

OBJECTIVES OF WORK PERMITS

The objectives of work permits are to eliminate the risk of injuries, occupational diseases, fires, and other undesired events occurring during maintenance, repair, or other work by specifying certain actions, sequence of actions, equipment, and controls. There are many types of work permits within the safety management system.

Examples of Work Permits are as follows:

- Hot work permit
- Electrical or mechanical work permit

- Confined space work permit
- Work at height permit
- Diving permit
- Excavation permit
- Contractors work permit
- Work with explosives
- Hazardous substance permit, etc.

Hot Work Permit

A *hot work permit* is issued for carrying out work in an explosive area or area where the fire risks are high. This would include hack-sawing, hammering, drilling, or filing. Open flame work in an explosive area would include tasks such as lead burning, grinding, arc welding, soldering, and oxyacetylene cutting or welding.

Electrical or Mechanical Work Permits

Electrical or mechanical work permits would include permits to work on electrical installations or mechanical installations. Work permits could be either electrical isolation permits or non-electrical physical isolation permits. This type of permit would include closing and locking valves, inserting slip plates, and physically disconnecting pipelines, or blanking off live ends.

Confined Space Entry Permits

Confined space entry permits are used for vessel entry or whenever work is to be carried out in underground passages, pipes, ducts, openings under columns, sewers and gutters, and pits or canals. These permits would also be required when entering into fermentation tanks, metal storage tanks, autoclaves, and other vessels. Any area defined as a confined space would require such a permit which is often a legal requirement.

Work at Height Permit

A *work at height permit* would be issued for people who work at the top of tall stacks, on high roofs, and in any elevated position, above 6 feet (2 m), and where the chances of falling are high.

Diving Permit

Another type of permit is an *underwater diving permit*, which would be issued for specialized applications, when work has to be carried out underwater. The two types of diving permits are a sea diving permit and a fresh water diving permit.

Excavation Permit

An *excavation permit* is normally required for any excavation where there is a chance of encountering underground buried services or other obstacles. Normally this permit would require plans to be consulted before the dig, or for some form of scanning to be done to ascertain the position of underground pipes, cables, etc.

Other Permits

There are many types of permits which could be used to ensure the safety of the employee and process, and although the specific permits and requirements may differ from company to company and site to site, they are a vital component of any safety management system.

A formal system of issuing and canceling permits should be part of the permit process. Trained issuers and trained permit receivers should be identified and retrained annually. The permit issued should be displayed at the job site for inspection. The use of permits and the permit system must be tested during the safety system inspections and audits.

ORGANIZATION RISK MANAGEMENT (S1, E1.8)

Risk management considers probabilities and outcome severity of undesired events occurring within the organization. This enables action plans to be put in place to reduce the likelihood of losses occurring. It combines the function of planning, organizing, leading, and controlling of the activities of an organization, so as to minimize the probability of the occurrence of, and the adverse effects of accidental losses. The risks presented by the operations form the driving factors for the safety system controls in the form of processes, programs, activities, standards, balances, checks, management and employee actions.

HAZARD IDENTIFICATION

Identifying hazards and the hazard burden within the organization is the first step in the Hazard Identification and Risk Assessment (HIRA) process. A number of hazard identification methods have already been discussed, and as with other safety processes, these are ongoing procedures.

RISK ASSESSMENT

To help determine the potential probability, frequency and severity of loss which could be caused by hazards, the technique of risk assessment is a major element in the safety system. The steps of risk management are as follows:

- Hazard identification
- Risk analysis
- Risk evaluation
- Risk control
- Re-assessment

The risk management process is to:

- Identify all the pure risks within the organization and which are connected to the operation (*hazard identification*).
- Do a thorough analysis of the risks taking into consideration the frequency, probability and severity of consequences (*risk analysis*).

- Implement the best techniques for risk reduction (*risk evaluation*).
- Deal with the risk where possible (*risk control*).
- Monitor and re-evaluate on an ongoing basis (*re-assessment*).

Daily Task Risk Assessment

Ongoing hazard identification and risk assessments are components of a safety management system, and can be cascaded down to the lowest level in the organization by means of employee and team daily risk assessments. The on-site daily, or task risk assessment, is a checklist or questionnaire that is completed by the work team leader on site before the task is undertaken. It calls for an assessment of *what could happen* during the task and possible outcomes. A simple risk matrix is completed based on these questions and if the risk is high, action is taken before the job starts. If the risk is medium, work can start while corrective action is taken, and if the risk is acceptable, the work can begin (Figure 8.2).

Daily or task risk assessment

Before each work task ask the following questions:
PROBABILITY = *What can happen here?* (Injury, damage, fire, etc.)
SEVERITY = *If it happens, how bad will it be?* (Death, injury, damage)

TASK...
Rank each question as Low (1) Medium (2) Medium-High (3) High (4)

Probability of accident		Low	Medium	Medium-high	High
	High	4	8	12	16
	Medium-high	3	6	9	12
	Medium	2	4	6	8
	Low	1	2	3	4

How bad could it be?

12–16 Stop work immediately and fix the unsafe situation
4–9 Fix the unsafe situation and continue work
If the work is accepted as safe – proceed

Action taken:...
....

Safe to work after action **YES** ☐   **NO** ☐   (Stop job)

Name...............................Signature............................Date................

FIGURE 8.2   A daily or task risk assessment.

WRITTEN SAFE WORK PROCEDURES (S1, E1.9)

Job Safe Practices (JSPs) or Safe Operating Procedures (SOPs), are required in order to reduce the possibility of risk to employees and damage to equipment while performing work tasks, specifically critical tasks. By following these set procedures, risk to the operator is minimized. Written procedures, properly used, serve as a base from which tasks can be continually modified to improve work methods and employee safety.

OBJECTIVE

The objective of critical task identification is to identify tasks (segments of work) that have high potential for loss if not carried out in a specific manner, and in a special sequence, and to reduce accident potential by identifying and instituting controls for the hazards presented in each step of each task.

This is achieved by analyzing the task, breaking it down into steps, identifying the risks for each step, and then establishing correct task methods with inbuilt controls, to mitigate or avoid the risks identified. This process is commonly referred to as Job Safety Analysis (JSA).

The process will develop written job safe practices (JSPs) for critical tasks to be followed, that will identify and control hazards that might not otherwise have been identified or highlighted. These will improve the safety, productivity, and quality of tasks, and improve work methods.

PLANNED JOB OBSERVATION (S1, E1.10)

Job Safe Practices, (JSPs) should be made available and used in the areas or departments where that particular critical task (job) is carried out. Employees who carry out these critical tasks should be trained in the JSP and must be supervised to follow the procedure without leaving out any steps, or otherwise deviating from the procedure. To ensure that the critical tasks are being carried out according to the JSP, the technique of job observation is used to observe the task while following it step-by-step on the JSP.

OBJECTIVE

The technique of planned job observation is used to observe the task being carried out while following it step-by-step on the JSP. The task observation involves observing the person, or team, carrying out a critical task according to the JSP. The main purpose of the observation is to ensure that the JSPs are being followed, that no critical steps are being left out, and that no short cuts, which can lead to injury-causing accidents, are taken.

PLANNED JOB OBSERVATION (PJO)

A planned job observation is a job observation that has been pre-planned on a schedule. For example, this could be every three months, once a year, or once a week.

Informal (not scheduled) job observations can also take place, and in some cases not all steps need necessarily be observed. Informal job observation forms part of the ongoing observation program.

PRIORITY

In choosing which critical task to observe first, the priority are those tasks that are high on the list of criticality as indicated by the critical task identification exercise. New and unusual tasks identified by legislation to be inherently dangerous should also receive priority for observations. It is important that the observations are carried out before accidents occur and not merely as post-accident activities.

SAFETY INSPECTIONS (S1, E1.11)

A core element of the safety system is safety inspections.

OBJECTIVE

The objective of this element is to ensure that a regular and structured inspection system is in place and working to identify unsafe conditions, high-risk acts, and non-conformances to safety standards so that corrective measures can be taken to reduce risks and prevent accidental loss.

The inspection standards and procedures should apply to all the organization's (and affiliate companies) workplaces and sites and include all areas, equipment and machinery within the company. This standard should be applied to contractors and their employees.

The summarized standard could read as follows:

The manager, supervisor, or contractor (Responsible Person) will ensure that all the work areas, buildings and installations under their control are subject to a formal safety inspection on a monthly basis as a minimum, where practicable.

SAFETY SUGGESTION SCHEMES (S1, E1.12)

OBJECTIVE

The objective of the safety suggestion scheme is to encourage and reward employees who identify hazards and risks in the company, and who suggest feasible solutions that eliminate or minimize such hazards, and also those who pose other safety suggestions which will help improve safety at the company. All non-supervisory employees, supervisory staff below manager level, casual employees, and vocational students should be eligible to make safety suggestions.

A successful suggestion accepted for implementation could be awarded based on the elimination or minimization of the hazard, or risk, defined as follows:

- *Major* (Award A) is given out for suggestions that eliminate or minimize a condition or practice likely to cause loss of life or body part, permanent disability, and or extensive loss of structure, equipment or material (Class A Hazard).

- *Serious* (Award B) is given out for suggestions that eliminate or minimize a condition or practice likely to cause serious injury or illness, resulting in temporary disability, or property damage that is disruptive but not extensive (Class B Hazard).
- *Minor* (Award C) is given out for suggestions that eliminate or minimize a condition or practice likely to cause minor, non-disabling injury or illness or non-disruptive damage (Class C Hazard).
- *General* (Award D) is given out for a general safety suggestion which will contribute toward further enhancement of safety within the organization.

EMPLOYEE JOB SPECIFICATIONS (S1, E1.13)

Another important safety management system element is the specifying of physical and cognitive requirements for each position within the company. This is especially important for employees who are exposed to risks while carrying out their normal duties, such as employees who work at height, with chemicals, work underground, drive equipment, or operate machinery, to mention a few. Getting the right person in the right position reduces the risk of that person being involved in an accident because they were physically or cognitively unsuitable for the position. All job descriptions should include the required mental and physical skills needed for the position as well as the ergonomic requirements of the work.

MEDICAL EXAMINATIONS (S1, E1.14)

Pre-employment medical examinations should be a part of the safety and health management system. This will help to establish a baseline health profile of all employees, which can be used to determine any deterioration of their health, as a result of the work they will be doing. Hearing acuity tests are done to establish hearing threshold levels, which are compared with up-to-date measurement, to ensure no hearing loss is being experienced as a result of sound levels in the workplace.

Employees who work with chemicals, hazardous substances, or who are otherwise exposed to other environmental hazards should undergo regular medical examinations to detect any adverse effects from the work environment. Dependent on the risk that employees are exposed to, some regular medical examinations may be more frequent than others.

OFF-THE-JOB SAFETY (S1, E1.15)

Off-the-job safety includes home, sport, and recreation safety as well as road safety and organizations should consider the safety of its employees and dependents, important both at work and off-the-job.

OBJECTIVE

The objective of this element is to promote the importance of off-the-job health and safety among employees, their families, and dependents in an effort to reduce accidental losses occurring away from the workplace.

This element of the safety system will involve the promotion of safety away from work so that the safety habits learned and practiced at the workplace are carried over into off-the-job activities. Influencing and informing the families about the hazards found in and around the home will help identify risks to children and loved ones, and help promote safety awareness away from the workplace as well. Posters, banners, and other visual material can highlight safety messages concerning the following:

- Home safety
- Water safety
- Road safety
- Fire safety
- Chemical safety
- Electrical safety
- Basic first aid, etc.

Off-the-job safety competitions, quizzes, and activities should encourage participation and involvement of all family members at such functions. Safety DVDs on home safety should be shown at the safety committee meetings at least quarterly. Safety notice boards could carry at least one off-the-job safety article, tip, or checklist each month. When available, safety posters depicting off-the-job safety themes should be displayed.

OFF-THE-JOB ACCIDENT AND INJURY REPORTING

Employees should be encouraged to report off-the-job accidents and injuries so that trends can be established as to common causes. This information can then highlight major off-the-job risk areas that can, in turn, be conveyed to other employees in an effort to avoid similar recurring accidents happening. During incident recall sessions, employees should be encouraged to report high potential near miss events that occurred to them or their family members.

SAFETY NEWSLETTERS (S1, E1.16)

Safety promotion is very valuable and can influence attitudes and behavior patterns. The promotion of safety and the safety system through consistent distribution and display of safety posters, slogans, banners, and safety DVDs, is an integral part of a safety system. The safety department could be responsible for producing a monthly safety newsletter, and for publicity in the company magazines, or publications. They should also maintain and control the distribution of slogans, banners, and safety DVDs.

SAFETY AND HEALTH REPRESENTATIVES (S1, E1.17)

The nomination, selection, and appointment of Safety Representatives is to ensure that sufficient and competent Safety Representatives are appointed in all work areas to promote safety and to conduct monthly inspections of those work areas.

Their appointment obtains participation and involvement in the safety system from the shop floor. In many countries this is a regulatory safety requirement.

Safety Representatives are appointed to identify hazards within their work areas. Since they are the employees most familiar with their work areas, they are therefore the best people to be able to identify hazards in the area. High-risk acts and conditions are the immediate causes of accidents, and if identified, reported, and eliminated, can lead to a reduction of risk in the workplace. Safety Representatives also assist in maintaining the momentum of the safety system due to their ongoing inspections, and liaison with employees, and their departmental heads.

Safety Representatives:

- Act as the eyes and ears for management by identifying and reporting hazards.
- Assist and participate in the ongoing company safety system.
- Assist in the promotion of safety to ensure a safer work area for all.
- Act as liaison between various employees and also between the employees and different levels of management.

SAFETY MANAGEMENT SYSTEM AUDITS (S1, E1.18)

OBJECTIVE

The objective of this element is to ensure that both internal and external audits of the safety management system are carried out at prescribed intervals, in all sectors of the organization to measure the degree of conformance to standards, and to evaluate the safety effort.

During the year, formal internal audits should be conducted to measure the implementation and effectiveness of the safety system and to identify non-conformances with standards. The organization should be audited every 6 months by internal auditors, and annually by external third party safety system auditors. All elements of the safety management system are considered during audits.

For audit purposes the following facilities and areas of the organization should be included in the audits:

- All work areas
- Storage yards
- Office blocks
- Reception areas
- Recreation facilities
- Hygiene facilities
- On-site accommodation
- Workshops
- Stores
- Contractor work sites
- Off-site premises and work areas belonging to the organization, etc.

AUDIT PROTOCOLS

Audit protocols are guidelines used for the audit, and should cover all the requirements and standards of the safety system. The protocol prompts the auditor as to what must be looked for during the inspection, what systems need to be tested, questions to be asked, and documented verification to be scrutinized. One standard protocol should be used for the entire organization's audit. If the audit is against a Guideline, then the audit protocol must correspond to the requirements of the Guideline. If it is a legal compliance audit, the protocol must cover all the regulatory requirements and regulations (Figure 8.3).

INTERNAL SELF-AUDITS

Self-audits are internal and do not result in outside recognition or accreditation by any authorizing body. They encompass audits of each facility within the organization and result in internal recognition and scoring of the organization. They are part of the continual improvement process.

EXTERNAL THIRD PARTY AUDITS (S1, E1.19)

External audits are carried out by an external agency or body and result in external scoring, accreditation, or recognition. They encompass the entire organization's facilities and premises, including all sites, and result in an evaluation of the company as a whole. During the audits, the auditors will inspect the physical work area and conditions, and then review the safety management system documented verification.

SAFETY PUBLICITY BOARDS (S1, E1.20)

OBJECTIVE

The objective of erecting safety publicity boards is to promote safety and health awareness and the safety management system by publicizing safety achievements, injury experience information, and other safety and health information. This is done by means of safety and health publicity boards erected throughout work areas, training centers, and office complexes.

The publicity board should contain information relating to the safety management system including the following:

- Local injury statistics
- Names of Safety Representatives
- Name and location of First Responder
- Safety pictures and awards
- Emergency procedures
- Emergency phone numbers
- Incident recall information

ELEMENT	PTS	PTS (MAX)	WHAT TO LOOK FOR	QUESTIONS THAT COULD BE ASKED	VERIFICATION
2.11 LOCKOUT SYSTEM AND USAGE					
2.11.1 WRITTEN PROCEDURE AVAILABLE AND APPLIED					
2.11.1.1 Are there task specific procedures available?		5	Is there a written lock-out procedure? Is the procedure valid?	Call for the procedure and check that it covers not only electricity but also pipelines, steam driven equipment, air equipment and other sources of energy, if applicable. Is the procedure clear?	Are the applicable work crews aware of the procedure? There MUST be a LOCK and a TAG always.
2.11.1.2 Are locks and tags used? Does the hardware meet the company standard?		5			
2.11.1.3 Are all affected staff trained?		5		Proof of training	
2.11.1.4 Are tags and locks used correctly?		5	Are the locks identified or the user identified?	Follow the system	During the inspection the system should be tested by endeavoring to find maintenance crews at work and checking to see whether or not the energy source has been locked out.
				Inspect locks on site	
2.11.2 CAN ALL EQUIPMENT BE ISOLATED AND LOCKED?					
2.11.2.1 The ideal is for all sources of energy to be lockable at the point of control/operation. The minimum to strive for is that all sources of energy can be individually isolated and locked at some point in the supply circuit to render it inoperative.		5	Do a random test of any piece of electrical equipment and ask the same question during the inspection. If other sources of energy, is there a lockout procedure?	Visual	Can all high voltage and low voltage circuits and critical valves, taps, etc. be locked out? Do they have a lockout box with various attachments?
2.11.3.1 Are all main switches accessible even when circuits are locked out?		5	Removing the wires behind the panel is not considered a lockout system nor does removing fuses alone. The lock-out system should give complete coverage as mentioned above.	Visual	If the doors to the distribution board are locked, can one still get access to the main switch?
TOTAL		35			

FIGURE 8.3   Example of one element of an Audit Protocol (Shown with scores 1–5 for each minimum standard detail).

- Safety and Health Policy
- Legal safety notices
- Workers compensation information, etc.

PUBLICITY, BULLETINS, NEWSLETTERS, SAFETY DVDS, ETC. (S1, E1.21)

Safety needs good publicity as it is normally only after a downgrading event that safety is highlighted. The safety system must be advertised and boosted by frequent informative emails, newsletters, and website postings. As much information about the system should be shared with all the organization's employees and other interested parties. Safety information cards for visitors should be printed and issued to them.

MEETING PROTOCOL

All business meetings and other meetings should start with the agenda item "safety," and a brief safety discussion should take place before the meeting continues with its normal agenda. This should form part of the organization's safety culture. Every meeting or gathering where visitors or strangers to the company are present should commence with the indication of the position of the escape routes from the venue and procedure to evacuate in case of an emergency.

SAFETY VISUAL MEDIA (DVDs AND ELECTRONIC VIDEOS)

There are many safety training and information DVDs, or electronic visual media available which can be used to reinforce the safety message and give up-to-date safety information. Visual media is the way most people communicate, and using such media to promote safety should form part of the safety publicity campaign. Institutions often compile a short video on their safety system highlights to use as part of the safety induction, as well as for the audit opening conference.

SAFETY COMPETITIONS (S1, E1.22)

Based on the management principle of recognition, and taking into account people are competitive by nature, safety competitions help engender enthusiasm and reward. Competitions must not be based on injury or severity rates, or any other measure of loss, but rather on controllable and manageable indexes such as internal audit scores, number and quality of near miss incidents reported, number of observations reported and rectified, departmental employees trained, or suggestions submitted within a certain period. Competitions help maintain safety motivation and generate enthusiasm among all levels of employees. They also measure the degree of the involvement of employees in the safety system.

Competitions could include good housekeeping competitions, the best Safety Representative, safety personality of the year, and other competitions that recognize excellence in safety.

The safety department is normally responsible for planning and arranging company-wide competitions on an annual basis, and for arranging suitable publicity for the competition and the winners. They could also be responsible for arranging other competitions within the organization on a regular basis, and for the promotion of these competitions and resultant publicity.

TOOLBOX TALKS, SAFETY BRIEFINGS, ETC. (S1, E1.23)

Short talks and discussions before the work commences is one way of communicating safety messages and warnings to work groups. Some organizations refer to them as tailgate meetings, or toolbox talks. Each discussion should include what work is going to be performed and what the hazards and precautions are. Reminders about the correct personal protective equipment and tools for the job form part of this short discussion. In some cases, a specific topic could be discussed as per a schedule. A sign-in sheet, which all present should sign, should be circulated at each meeting to verify acknowledgment and attendance.

SAFETY HOUR AND STAND-DOWNS

Safety hour is a system where once a week, a certain hour of the day is dedicated for all levels of managers to go down to the shop floor to have a discussion on safety with their employees.

Safety stand-downs are when the entire organization comes to a halt and multiple briefing sessions are held at all levels. This traditionally occurs after a major event or a fatal injury, but can be very beneficial if held without being triggered by a devastating event.

SAFETY SPECIFICATIONS: PURCHASING AND ENGINEERING CONTROL, NEW PLANT, AND CONTRACTORS (S1, E1.24)

When purchasing new equipment, machinery or plant, the organization should always consider the safety requirements of such purchases. There are a number of considerations that should be taken into account before new equipment or plant is acquired and put into use. Are machines correctly guarded, wired, and labeled? Have lockout features been incorporated, and is the equipment correctly color coded? A management of change process should be used when any new equipment, plant, process, or project is considered.

DUE DILIGENCE

Due diligence is the process through which a potential acquirer evaluates a new process, new piece of equipment or plant, before signing acceptance thereof. The theory behind due diligence is that performing this type of review contributes significantly to informed decision making by enhancing the amount and quality of information available to the organization purchasing or accepting the new plant or equipment. Safety reviews of new sections of plant or a new process line should be undertaken during and after construction to ensure safeguards have been correctly built-in. Before handover of equipment and plant, a thorough inspection, against a checklist, should be done to ensure the equipment meets the legal and organization's safety standards and meets the requirements of the management of change document.

CONTRACTOR SAFETY

One of the biggest challenges to an organization's safety system, is the control of contractors. Contractors should be regarded as part and parcel of the organization and its safety system. The safety requirements of the organization must be included in the pre-bid document, and a contract bid (invitation to tender) package must be sent to each contractor explaining the safety requirements and standards expected from the successful contractor. The organization's safety policy, the employee safety rule book, safety standards, safety requirements, and other documents should be in this package.

PRE-BID DOCUMENT

The contractor should complete a safety history questionnaire which accompanies their bid (quotation) for the work. On the questionnaire the contractor should provide:

- Their safety and health policy signed by the CEO.
- Name of senior manager responsible for site safety.
- Safety Coordinator allocated to the project (qualifications, affiliations, and experience in safety).
- Safety management system outline and standards.
- Their safety rule book.
- Safety training program.
- Work permit training.
- Name of First Responder.
- Names of Safety Representatives.
- Emergency plans, etc.

Contractors should agree to participate in the organization's safety committee meetings. The organization's contractor representative should also attend the company's safety meetings. Regular site inspections should take place and the contractor must undertake to rectify any hazards noted during these inspections.

SAFETY ORIENTATION

All contractors and subcontractors should attend site safety orientation training and be certified to enter the site. This would include vendors or visitors who spend more than 8 hours on site. Regular safety refresher training must take place.

SAFETY RULE BOOK (S1, E1.25)

A safety rule book is a small booklet which contains pertinent safety and health information for the employee. It contains a copy of the safety policy statement as well as the general safety and health requirements of the company. The rule book should carry a safety message from the chief executive officer, and is given to all employees on commencement of work at the company. This includes all contractors and long-term visitors, vendors, etc.

The book should form part of the safety induction and orientation training and each recipient of the book should sign acknowledgment of receipt of the book and its contents, acknowledging that they have been trained in the rules contained therein, and agree to abide by these rules. A short quiz on the content can be included as part of the induction training. To make the safety message attractive, the book should be colorful and where possible pictures, sketches, and diagrams should be used.

SAFETY REFERENCE LIBRARY (S1, E1.26)

A safety and health reference library contains publications, articles, and reference material pertinent to safety and health, as well as copies of safety legislation and regulations applicable to the organization. Copies of safety visual media form part of the library's collection. The library is more than likely to be in both soft and hard copy, and employees and other interested parties should have access to the material.

SHARING INFORMATION

Copies of the applicable safety law, Guidelines, safety system standards, check-lists, audit protocols, etc., can be stored in the safety library, in hard copy or electronic format. Books on safety and health should be purchased regularly and placed in the library for reading. A system of booking out hard copy documents must be in place and the information should be shared with all. An index to the material could be circulated so that sources for safety training workshops and toolbox talks can be easily obtained.

PUBLIC SAFETY (S1, E1.27)

The safety system should also consider the safety, health, and welfare of its clients, customers, and general public. This could be the end user of its products, or the effect its processes may have on the environment and the people living or working in the immediate facility. A mature safety management system extends beyond the organization and applies its principles and practices to others that may be impacted by it. Some organizations deal mainly with the public and therefore would have numerous processes and safety systems focused on protecting the people who use its services or products. A separate public safety system may be needed in some cases. Employees, contractors, and subcontractors and members of the public should all be considered as part of the safety system.

ANNUAL REPORT—SAFETY AND HEALTH (S1, E1.28)

OBJECTIVE

The objective of this element is to gather, produce, and disseminate safety information, facts and progress reports, in the form of various reports to ensure that safety and health achievements and progress are publicized at all levels. Interested parties would include shareholders, customers, international partners, members of the public, and others. A safety and health report should be submitted to the Board of Directors for discussion at the annual board meeting and should be included in the Annual Report.

PREPARATION

This report is prepared by the safety department in conjunction with management, and it contains both leading and tailing performance indicators. Safety and health achievements are also highlighted in the report. It could include the progressive lost time injury incidence rate and the number of injuries recorded during the period. Safety performance indicator achievements, safety system progress, and audit results are also included in the board papers as well as other safety highlights from the preceding 12 months. This information is summarized and reproduced in the company annual report and is also posted on the organization's website.

SAFETY SYSTEM DOCUMENTATION CONTROL (S1, E1.29)

Safety documentation would consist of all the policies, standards, procedures, checklists, and documented evidence appertaining to the safety system. Safety standards need to be updated, modified, and reviewed on a regular basis as part of the continual improvement of the safety system. These documents (hard or soft copies) need to be controlled so that new revisions replace the old ones and all holders of the documents, and those who access the documents, are informed of the updates, modification, and changes. All changes to the documentation must be approved by the established approval hierarchy.

A weak point in a safety management system could be that old policies are still in circulation despite the fact that they have already been updated or modified. A system of withdrawing the old document and replacing it with the updated one needs to be incorporated as part of the safety system's processes.

All verification of safety system documentation should be numbered according to the safety system numbering sequence, dated, and signed by the appropriate official. Electronic signatures are acceptable. Documents must be valid and current. For audit purposes, verification documentation should be retained for at least 12 months.

CONTINUAL IMPROVEMENT (S1, E1.30)

A safety management system needs to be continually improved or else it stands the chance of stagnating. New developments and initiative should be introduced to improve the system and keep employees engaged in its processes. Planned improvements should be set as objectives with completion dates. Records of improvements should be kept for reference. Ongoing inspections, reviews, and audits will provide scope for further system enhancements. Keeping the system and its elements evergreen will be a task for the leadership team supported by a strong and proactive safety department.

9 Electrical, Mechanical, and Personal Safeguarding—Part 1

EXAMPLE SMS: SECTION 2

Section 2 of the Example Safety Management System (Example SMS) consists of 23 Elements. The first 10 elements will be discussed in this chapter and the last 13 elements in Chapter 10 (Figure 9.1).

INTRODUCTION

Sources of energy need to be controlled to prevent accidental contact and subsequent loss. Section 2 of the Example SMS lists the most common elements dealing with the guarding of machinery, electrical components, and the worker (once all other means of protection have been applied).

Under this section, there are at least 23 elements containing systems, sub-systems, programs, processes and actions, which must be part of the safety management system. Each element should have a written standard which indicates the purpose of the standard, allocated responsibility and accountability, actions, and measurable performance criteria.

PORTABLE ELECTRICAL EQUIPMENT (SECTION 2, ELEMENT 2.1) (S2, E2.1)

Portable electrical equipment refers to all electrical equipment fed through a plug and a flexible cord. The inspection of this equipment on a regular basis will ensure that unsafe equipment is not used. Each area manager, supervisor, and contractor should be responsible for the portable electrical equipment under their control and must ensure that defective equipment is tagged and removed from the site for repair. Employees are required to report any defective portable electrical equipment. Depending on the risk (usage, exposure, etc.), the equipment should be inspected as required and these inspections should be documented.

REQUIREMENTS

Each supervisor should ensure all portable electrical equipment in use within his or her responsible area, is inspected before use and on a regular basis. This includes portable electrical equipment used in offices, for example, kettles, microwave ovens, etc.

Number	Element Title	Number	Element Title
2.1	Portable Electrical Equipment	2.13	Personal Protective Equipment (PPE)
2.2	Ground-fault Interrupters	2.14	Fall Protection
2.3	Fixed Electrical Installation	2.15	Ergonomics
2.4	Machine Guarding	2.16	Hearing Conservation
2.5	Ladders, Stairs, Walkways and Scaffolding	2.17	Occupational Stress Management Program
2.6	Safety Signs	2.18	Respiratory Protection Program and Equip.
2.7	Hazardous Substance Control	2.19	Blood Borne Pathogens Program
2.8	Welding and Cutting Safety	2.20	Food Safety program
2.9	Hand Tools	2.21	Lifting Equipment, Gear and Records
2.10	Powered Hand Tools	2.22	Compressed Gas Cylinders: Pressure Vessels
2.11	Lockout, Tag-out and Tryout (Energy Control)	2.23	Motorized Equipment: Checklist, Licensing
2.12	Labelling of Switches, Controllers, Isolators, etc.		

FIGURE 9.1   Section 2 of the Example SMS (23 Elements).

Equipment in use in offices such as computers, printers, etc., could be included in the general office safety inspections. Any new equipment should be inspected prior to putting it to use for the first time. The person performing the inspections must be competent to perform the inspections by virtue of their trade, or training in the inspection of portable electrical equipment.

GROUND-FAULT INTERRUPTERS (S2, E2.2)

Objective

The objective of this element standard is to set the company requirements for installation, use, and testing of Earth Leakage Protection (ELP) or Ground-Fault Circuit Interrupt (GFCI) protection in 110/240 Volt Alternating Current (110/240 VAC) circuits. This safety system standard should be designed to meet or exceed applicable international and local regulations. This standard should apply to all company properties, and any facilities maintained by work groups from these properties.

GFCIs should be installed in:

- Receptacles in any area, where water is frequently used in washing down floors for cleaning or dust suppression.
- Receptacles in any area where water commonly drains to and remains standing on working surfaces.
- Receptacles supplying water coolers.
- Receptacles within 6 ft. (1.8 m) of open vats/pools/cells of liquid in which portable electrical equipment could be immersed.

PORTABLE UNITS

In areas where installed GFCI protection is not available, it may still be required under some circumstances. In these situations, portable units may be used. A partial list, as example, includes the following situations:

- Temporary lighting and/or power installed in any enclosed metal vessel.
- Working in any pit/sump/cellar where standing water has collected.
- Using extension cords to supply power to areas above, from receptacles not GFCI protected.

FIXED ELECTRICAL INSTALLATION (S2, E2.3)

OBJECTIVE

The objective of this element is to provide for the protection of people and plant from damage due to the effects of electricity, by establishing minimum safety requirements for electrical installations. This is an important element of the safety system as in 2014, *Electrical, Wiring Methods* was ranked as one of OSHA's top ten most cited violations with nearly 3,000 violations issued. *Electrical, General Requirements* was also ranked in the top ten with 2,427 violations. (OSHA website, 2015b)

STANDARD

Electrical installations, and the practice surrounding them, must conform to the requirements of local electrical wiring and installation regulations, recognized international standards, as well as relevant company policies. Electrical installation works and equipment installed and maintained by an organization must meet the requirements of legislated electrical installation standards.

In situations where a site standard electrical specification exists that is relevant to a design or installation, the requirements of that specification are to be met. Site standard specifications must be equivalent to, or exceed, the requirements of recognized international standards. Only Competent Persons should carry out electrical work.

Electrical equipment and switchgear must be suitably enclosed and sign-posted at all times, to prevent access to live conductors by employees, other than those authorized persons or technicians issued with a special tool or key.

Authorized electrical workers, and those assisting them, must have undergone training in techniques of sub-station rescue and resuscitation. This training ensures that they are capable of providing competent assistance in the rescue of a victim in the event of an electrical accident. General electrical installations should be inspected annually for correct grounding and polarity.

ELECTRICAL ARC FLASH PROTECTION PROGRAM

Dependent on the nature of the workplace, a major component of the electrical safety plan would be the identification of situations where electrical arc flash could occur and injure employees. Once areas and situations are identified, risk assessments and labeling should take place, followed by employee training and the supply and use of appropriate arc flash PPE and clothing.

MACHINE GUARDING (S2, E2.4)

A machine guard is a device that prevents limbs from contacting the dangerous moving parts of machinery, and guarding means effectively preventing people from coming into contact with the moving parts of machinery or other equipment that could injure them.

Unguarded or inadequately guarded machines cause a large number of serious injuries. These injuries are either permanent or severe in their nature. More than 2,500 violations concerning machine guarding were recorded by OSHA in 2014, making it one of the top ten most cited violations for that year. (OSHA website, 2015b)

Machine guarding is applicable at most workplaces, at homes and even in shops and offices. A rule of thumb is that if a person should stumble and fall with outstretched arms, could they be injured in a machine or unguarded pinch point? Machine guards should be fitted wherever rotating or operating machinery is within normal reach.

Any moving, rotating, or vibrating part of any machine that could injure a person should be guarded. This would include the following: projecting shaft ends, transmission belts, band saws, band knives, planning machinery, shears, guillotines, and presses, and any other device that poses a hazard if unguarded.

Enclosing means guarding by means of physical barriers that are mounted on a machine in an effort to prevent access to the hazardous parts. Fencing means erecting a fence or rail which restricts access to a machine.

CLASSES OF MACHINE GUARDS

There are two basic classes of machine guards. They are *transmission* and *point-of-operation* guards.

- *Transmission* guards are guards, which guard all mechanical components including gears, cams, shafts, pulleys, belts, and rods that transmit energy and motion from a source of power to a point of operation.
- *Point-of-operation* guards, are guards that effectively shield the area on a machine where the material is positioned for processing and where an exchange of energy takes place.

Types of Machine Guards include the following:

- A *fixed guard* is preferable and should be used in all cases where possible. It prevents access to the danger areas at all times and is normally a part of the machine.
- *Interlocking guards* are either mechanical, electrical or pneumatic, and prevent the operation of the controls that sets the machine in motion until

the guard is in position. Removing or opening the guard locks the starting mechanism and the machine cannot be operated. An interlocking guard must guard the nip point before the machine can be operated. It should stay closed until the dangerous part is at rest, and it should prevent the operation of the machine if the interlocking device fails.

- An *automatic guard* functions independently of the operator and its action is repeated as long as the machine is in operation. Automatic guards are fitted where neither a fixed nor interlocking guard is practical. Examples of automatic guards are two-handed push buttons, pull back devices, and photoelectric devices.
- *Point-of-operation guards* shield the area where the nip point is generated.

LADDERS, STAIRS, WALKWAYS, AND SCAFFOLDING (S2, E2.5)

OBJECTIVE

The objective of this element is to reduce the risk of fall-related injuries by the inspection and control of ladders, stairs, walkways, platforms, and scaffolding.

LADDER SAFETY

Ladders are a quick, safe, easy and convenient way of gaining access to positions beyond our normal reach. Ladders are used extensively every day in industry, at home, underground in mines, and in many other applications.

Despite the progress made in the design and construction of ladders, they still tend to be misused, incorrectly used, and wrongly stored. Each year this results in injuries as a result of falls from ladders. Statistics also show a significant amount of fatal accidents as a result of falls from ladders. This could be due to either the faulty condition of the ladder, the faulty or incorrect use of the apparatus, or a high-risk act by the user of the ladder that caused him or her to fall to the ground and suffer an injury. Incorrect ladders and usage, resulted in ladder violations being one of the top ten OSHA citations of 2014. (OSHA website, 2015b)

RESPONSIBILITY AND ACCOUNTABILITY

Managers, supervisors, and contractors, should be responsible and accountable for ladders. They must ensure that all ladders, stairways, platforms, and scaffolds are inspected and tagged, and hazardous items repaired or removed from service. This includes the overseeing of the education and training of employees in the care and use of ladders, stairways, and scaffolds.

A ladder, stairway, and scaffolding safety program would consist of the following:

- Training in the safe use
- The need for ladders
- Purchasing
- Correct usage of ladders

- Storage of ladders
- Electrical ladder safety
- Ladder inspections
- Fixed ladders
- Stairways
- Scaffolding safety

Training

Safety Representatives and safety personnel should do the education and training of employees, and should also do the periodic inspection of ladders, walkways, stairs, and scaffolds. All employees are also responsible for the inspections of, and condition of ladders, stairs, walkways and scaffolds, the safe use thereof, and should not work on scaffolding during adverse weather which could put them at risk. The safe use of ladders should also be discussed at toolbox talks and safety briefings and should be included in the task risk assessment.

Need for Ladders

Where the need for a ladder can be eliminated by providing a fixed platform or other means of access, this would eliminate the risks associated with the use of ladders. Where this is not possible, a ladder safety program needs to be in place as part of the safety system.

Purchasing

The purchase of ladders should be done in conjunction with the purchasing department, the department requesting the ladder, and the safety department. Specifications should be drawn up concerning the type, strength and length of ladder, manufacture of ladder, and specifically the safety devices required on that ladder.

A pre-use inspection should be carried out on each new ladder before being put to use. At this inspection the ladder can be identified and numbered for the control system. The ladder should then be entered onto the ladder checklist or inspection tag attached to the ladder. Thereafter, the first inspection should be carried out.

Correct Usage

The prime rule for ladder safety is that the correct ladder must be used for the job for which it is intended. Studies of causes of ladder accidents invariably show that by far the majority of accidents are due to the ladder being used for a job for which it was not intended.

Extension ladders have a specific task, self-standing ladders are intended to be used as such, and fixed ladders against the side of chimneys, steeples, and industrial buildings are also only meant for access to the elevated place and not for any other purpose. The use of durable but strong material has led to ladders becoming lighter and consequently more portable and less of a lifting risk when being moved from position to position.

In certain countries, ladders have to be approved by an approval authority and a sticker attached containing certain basic safety rules as well as the maximum load rating for that ladder.

Storage of Ladders

Ladders should be stored away from the elements and should be stored in an area where they are easily accessible. Always store long ladders on their sides, supported at the bottom in properly designed racks. These racks should have sufficient support along the length of the ladder to avoid sagging.

Materials and heavy objects must not be stored on the ladder while it is in storage, as this could put excessive sideways loading on the ladder and weaken it. Aluminum ladders should be stored away from all chemicals, which may set up a corrosive process. Ladders should never be stored on the floor or laid down in work areas, as severe damage can occur should they be bumped or ridden over by fork trucks or other vehicles.

Electrical Ladder Safety

Aluminum ladders, though light, portable, and very strong are also conductors of electricity and care must be taken that they do not come into contact with overhead electrical wires or other open conductors such as overhead crane buss bars, etc. Out in a field, erecting aluminum ladders can be hazardous due to the presence of bare overhead conductors.

The placing of a ladder should be under the direct supervision of a competent person, who should check that aluminum and other conductive ladders are not used anywhere where there is a possibility of contact with electric current. Many electrical utility companies have banned the use of conductive ladders on their sites.

Ladder Inspections

All portable ladders should be checked before use and according to their usage (the frequency will be dependent on the risk). Each ladder should have an inspection tag attached for inspection purposes, or should be numbered and the numbers recorded in a register, listing the condition of the ladder as well as any comments. Portable ladders should be clearly identified and unsafe ladders should be tagged and removed from service immediately. When not in use, ladders should be stored in clearly demarcated ladder racks.

Fixed Ladders

Fixed ladders are just as important as any other ladder and they must be designed to withstand a single concentrated load of at least 200 lbs (90 kg). Fixed ladders should also be identified and inspected regularly.

A fixed ladder that exceeds 15 ft. (5 m) in height should be provided with a cage, which should extend from a point not more than 8 ft. (2.5 m) from the lower level of the ladder to at least 3 ft. (900 mm) above the top of the ladder. The cage should provide firm support along its entire length, for the back of a person, and no part of the cage should be more than 2 ft. (600 mm) distant from the plain of the rungs. (OSHA 1910.27 *Fixed Ladders*) (2015c)

Stairways

Stairways should be maintained in a good condition, free from defects and obstructions, and stairways and landings should not be used as storage areas. Stair design should conform to international building code standards. Handrails are needed if there are four or more risers.

Scaffolding

Scaffolding should be erected by a specialist or company that erects scaffolds. The scaffold must be inspected and passed before first use and again at regular intervals. Any defects noted should result in the scaffolding being closed and barricaded until defects are rectified. A visible red (no go) or green (go) tag is sometimes attached to the entry point of the scaffolding indicating its state of readiness after an inspection.

Where employees have to work in a location where there is a risk of falling more than 6 ft. (1.83 m), a proper platform with guard rails must be provided, or if it is impossible or impracticable to provide a safe working platform, complete personal fall arrest systems consisting of a full body harness, shock absorbing lanyard, and anchor point should be used.

SAFETY SIGNS (S5, E2.6)

The high-risk act of "failure to warn," is responsible for a number of injury-causing accidents every year. Correct signs can warn and help prevent accidents caused by the person being unaware of, or not being warned of the danger or precautions. Safety signs should be standardized throughout the plant so that employees become familiar with them.

Safety signs form an important part of any safety management system as their purpose is to inform about certain safety conditions. They also display information, warnings, and indicate what action must be taken in certain cases. Safety signs convey numerous messages and their purpose can be summarized by the following:

- Warn of dangers
- Inform about conditions
- Show where items are to be found
- State certain safety facts
- Indicate action to take
- Reinforce existing safety rules
- Enforce certain safety measures
- Prohibit certain actions
- Impart certain safety messages

Safety signs communicate with people in a work area. The communication can be in the form of a warning, a message prohibiting behavior, a message emphasizing a mandatory action, or they could give information and indicate locations and positioning of services.

SIGNAGE SURVEY

As part of the safety system, an initial safety signage survey (sign risk assessment) should be undertaken to determine what signs are needed and what signs need repair or are redundant. Signs should be clean, legible, and reflective where applicable. Safety signs should be included in the regular safety inspection checklists.

CATEGORIES OF SAFETY SIGNS

The three categories of safety signs are a sign which contains a printed message, a pictorial safety sign that depicts the message by means of a diagram or sketch, and a combination safety sign which uses both written message and a pictorial message.

Symbolic Safety Signs

In many countries, organizations with a multilingual workforce use symbolic safety signs which consist of four characteristics: a standard size, different shapes for different messages, different background colors, and a pictorial message without any writing.

In one symbolic sign there are four characteristics that differentiate it from other signs. The sign is of a particular size, color, and shape and has a clearly understood pictorial message. The four shapes are as follows:

- Triangular—for warning. (Yellow)
- Circular "annular" or ring—for prohibited. (Red)
- Circle (disc)—for mandatory messages. (Blue)
- Square—for information of direction. (Green)
- Square—for fire information. (Red)

There are four main colors used for symbolic signs:

- Yellow for warning.
- Red with a diagonal slash for prohibited.
- Blue for mandatory.
- Green for information of direction and safety equipment.
- Red for information of fire equipment.

Therefore, all warning signs are triangular with a yellow background, prohibited signs are red circles, mandatory signs are blue discs, information signs are green squares and fire information signs are square or rectangular with a red border.

Traffic Signs

Road signs and traffic signs are categorized as safety signs as they warn the motorist or pedestrian and give information. Traffic signs indicate where vehicles are to stop, where they are not to stop, and at what speed they must travel, and also give vital information to ensure the safe flow of traffic. As motorized transport is used in and around most industries and mines, these traffic signs are an important part of the safety information process. Where possible, the traffic signs used should conform to the signs used on public roads to ensure there is no confusion and no difference between the standard signs.

PPE Usage Signs

These signs are used in many instances to indicate what item of personal protective equipment must be worn in certain areas and during certain processes. When entering a work area the signs normally inform the visitor of the rules concerning personal protective equipment. This ensures that the person is warned and informed before entering the workplace. If signs were not used, the person would possibly enter the workplace and could suffer injury as a result of not wearing the required personal protective equipment.

Hazard Warning Signs

Signs are also used to warn workers of the hazards in workplaces and the hazards of the materials they are handling. The process hazards are also indicated by means of signs and where toxic or corrosive chemicals are used. Signs warn of the danger.

Directional and Information Signs

Signs also give information as to what and where. For instance, signs will indicate what type of fire equipment is situated and where. Signs also inform as to the direction of emergency exits, the position of first aid equipment, and also where telephones are located, etc. In an emergency, signs play an important role in indicating the positioning of emergency equipment, fire exits, escape routes, assembly points, first aid facilities, etc.

Signs to Promote Safety

Safety signs are also used to promote safety messages in a work place. To ensure there is an ongoing awareness of safety, safety signs constantly promote safety with their messages. They can be used to advertise a safety campaign or a safety goal that has been achieved.

When erected on the road leading out of the plant they can leave departing workers with a final safety message such as to fasten their vehicle safety belt.

HAZARDOUS SUBSTANCE CONTROL (S2, E2.7)

Hazardous substances are chemicals and other substances, which possess the potential to cause injury or disease to persons under certain circumstances and situations. They are substances that, due to their chemical properties, constitute a hazard.

Hazardous substances include the following:

- Explosives
- Oxidizing agents
- Flammables
- Corrosives
- Toxic substances
- Poisons
- Harmful or irritant substances
- Aerosols
- Radioactive emissions, etc.

DANGERS

The dangers associated with hazardous chemicals vary, and certain factors which determine the degree of hazard, are the following:

- The type of exposure.
- Whether it is continuance or intermittent exposure to more or one hazardous substance.
- Personal physical differences.
- Previous work exposure.
- Nature of the hazardous substance.

The degree of illness or injury is also dependent on the toxicity of the material and the effects may be classified as either systemic or local. Systemic effects are the affects produced by the chemical on a whole range of bodily functions, often far removed from the rate of entry into the body. Local effects are produced at the point of contact of the chemical with the body.

HAZARDOUS SUBSTANCE CONTROL SYSTEM

A structured system is needed for the control and use of hazardous substances at any workplace, and an example is as follows:

Step 1. Identify

All chemicals and hazardous substances in use should firstly be found, identified, and listed. This would involve reviewing the process, listing the substances used, recording the by-products and identifying what hazardous substances are being purchased by the organization.

The identification process will involve inspections of the work areas, and should include a process review to establish whether or not hazardous chemicals can either be eliminated from the process or replaced by less hazardous substances.

All areas should be covered during this initial inspection and each department could be asked to complete a hazardous substance list. This list would indicate all the substances used or stored by each department to enable a complete list of hazardous substances used on site to be compiled.

Step 2. Appoint a Coordinator

A hazardous substance coordinator should be appointed to coordinate the purchase, use, and disposal of all hazardous wastes. This coordinator should have a basic understanding of occupational health and he or she should be familiar with the applicable legislation and controls concerning hazardous substances.

Step 3. Record

Once all the hazardous substances on site have been identified, a master list must be prepared. This list will list all of the hazardous chemicals used, as well as their trade names, chemical compositions, and hazards.

Certain information on chemicals may need to be obtained from the supplier. The coordinator should contact suppliers asking them to provide material safety data sheets (MSDSs). A material safety data sheet (MSDS) is a sheet listing the characteristics, names, composition, and hazards of the substance as well as the necessary controls both for use, storage, and procedure in case of emergency.

Once this master list of hazardous substances has been compiled, the substances safety data sheets should be classified alphabetically and by known names as well. Many references are now available online for speedy reference. These websites are also updated on a regular basis.

It is important that material safety data sheets be kept in the division where the hazardous substances are used or stored. Departments that are involved in the disposal of hazardous substances, or in the handling of hazardous substance emergencies, should also have copies of the material safety data sheets.

Step 4. Controlled Purchasing

A system to ensure that hazardous substances are not purchased and used without notifying the hazardous substances control coordinator, should now be introduced. The purchasing department should complete a form when applying for the introduction of a new chemical and this form should state:

- The name of the product
- Where the product will be used
- The reason for introducing this product

These applications will then go via the coordinator who will look for less hazardous substitutes which may be available on the market. If no substitute is found and the chemical is approved, the necessary material safety data sheets will be compiled, and incorporated into the training, etc.

Step 5. Disposal

Correct written procedures should be drawn up for the disposal of hazardous substances and should be available. Liaison with the necessary authorities should be ongoing and compliance to legal requirements should be met at all times. Workers who are involved with the handling and disposal of hazardous substances should receive adequate training, both theoretical and on-the-job. The necessary and correct types of personal protective equipment should be supplied.

Step 6. Training

The required training should be given in the use, storage and disposal of hazardous substances. People working with the substances should receive priority training, and others who are involved in the storage, disposal, or emergency handling should receive adequate training in the hazards associated with the substances.

Step 7. Operating Procedures

Written operating procedures should be followed by any employee handling and working with hazardous substances. Training in these procedures should commence with induction training and ongoing training must be available to ensure that the procedures are being followed at all times.

If the hazardous substance and its procedure have been identified as a critical task, the necessary job safety analysis must be conducted. The safety rulebook should also include the basic safety procedures for handling, storage, and disposal of the hazardous substances on site.

Step 8. Emergency

There are various emergencies that can occur when working with and handling hazardous substances. Even storage of hazardous substances can result in certain emergencies. Written emergency procedures should be made available for employees to scrutinize. Ongoing training in the emergency procedures and regular emergency drills would ensure complete readiness in case of an emergency.

WELDING AND CUTTING SAFETY (S2, E2.8)

The safety management system should have identified all risks associated with welding, grinding, and cutting activities. By their very nature these activities are hazardous and should be risk assessed and the necessary controls should be put in place to reduce the hazards created by these processes. Items that the system should consider include: PPE, hot work permits, electrical safety, oxyacetylene gas safety, barriers and screens, and other control measures derived from the risk assessments.

HAND TOOLS (S2, E2.9)

Hand tools are items such as hammers, chisels, knives, saws, spanners, and wrenches, etc., and include other items such as barrows and carts, etc. Numerous injuries occur each year as a result of accidents caused by improper use, or poor condition of hand tools. The hand tool itself should be the correct type and should be in a safe condition. The person using the hand tool must use it correctly and these two factors combine for the safe use of hand tools.

HAND TOOL STANDARDS

The safety element standard for the safe selection and use of hand tools should include the use of the correct tool for the job in hand, prohibiting the use of make-shift or homemade hand tools, keeping hand tools in a good condition, correct tool storage, keeping tools in a safe place to ensure that they do not fall from overhead, correct training, and the regular inspection of hand tools.

Hand Tool Storage

Most users take pride in the way they store and display their hand tools. Often hand tools are hung up on pegboards or in a tool chest and shadow painting is used to mark the position of these tools. This practice leads to the instant recognition of a missing hand tool and also allows for easy return of the tool to its proper position. This improves the housekeeping in cabinets and cupboards and generates an awareness of order.

Training

Newcomers to the industry, learner technicians, and apprentices should be trained in the safe use of hand tools from the beginning of their work career. They should learn how to identify the correct tool for the correct job, and how to keep and store tools in a safe condition. Private hand tools used at work should also be part of the ongoing inspection system.

Inspections

Supervisors are responsible to inspect the condition of hand tools being used. The identification of the work being carried out should also be monitored on an ongoing basis and this will include the correct use of the correct tool. The hand tools could be included in the Safety and Health Representative monthly checklist to ensure that they are inspected monthly. Where necessary, special tools may be required and supervisors should check to ensure they are being used. Examples of these are non-spark tools used in areas where there is a high risk of explosion or fire.

POWERED HAND TOOLS (S2, E2.10)

These include powered and pneumatic hand tools which are powered by electricity, battery, compressed air, or small gas engines, etc. Also included are explosive powered tools.

There are numerous powered hand tools being used at the workplace which can also be hazardous. In the case of electrical powered tools, all the necessary safety checks should be in place to ensure that a GFCI unit protects the device. The polarity must be correct and the cord and switches should be in a good condition. Suitable eye and hearing protection should be worn, when and where necessary.

In the case of pneumatic powered hand tools, suitable eye protection should also be worn as well as hearing protection when necessary. Regular checks should be made on the source of compressed air to ensure that the necessary machine guards are in place and that the operating pressure and automatic cut-out switch and other safety devices do operate.

Explosive powered hand tools need to be controlled as required by regulatory standards which would cover aspects such as the following:

- Usage
- Training of operators
- Storage
- Control of cartridges, etc.

10 Electrical, Mechanical, and Personal Safeguarding—Part 2

LOCKOUT, TAGOUT, AND TRYOUT (ENERGY CONTROL) (SECTION 2, ELEMENT 2.11) (S2, E2.11)

The control of energy sources during maintenance or repair work is a vital element of the safety management system (SMS) (Figure 10.1). OSHA, during 2014, ranked this element as number 6 in the top ten violation list. (OSHA website, 2015b.)

STANDARD ESTABLISHED

A standard should be established to safeguard persons and property from hazards arising from all forms of stored energy. This should incorporate an energy control process that provides the utmost protection for employees servicing or working on equipment, during which the unexpected start-up of the equipment or a release of stored energy, could cause injury to the employees and/or damage to the property.

ISOLATE, LOCK, TAG, AND CHECK

The standard should state the basic provisions considered necessary for the safety of property and employees, before being allowed to work on portions of the system or equipment. Energy sources should be isolated in a safe manner by locking them out, tagging them and testing the isolated energy source.

The energy lockout, tag, and tryout procedures should be applicable for safely isolating all energy sources such as electrical, stored electrical, other stored energy, hydraulic, pneumatic, thermal, process gases, fluids, chemical, and mechanical sources of energy. The standard should cover all hazardous sources of energy, irrespective of the type and/or voltage.

ALL SOURCES OF ENERGY

Energy sources include the following: distribution power lines, control and operation panels in power plants, switches, disconnects, cables, feeders, circuits, water pressure, air pressure, other stored energy, gravity, etc., and associated equipment and devices which are sources of stored energy. Low voltage power lines and associated equipment in office complexes, buildings, and facilities should also be included as sources of energy that need to be locked out.

Number	Element Title	Number	Element Title
2.1	Portable Electrical Equipment	2.13	Personal Protective Equipment (PPE)
2.2	Ground-fault Interrupters	2.14	Fall Protection
2.3	Fixed Electrical Installation	2.15	Ergonomics
2.4	Machine Guarding	2.16	Hearing Conservation
2.5	Ladders, Stairs, Walkways and Scaffolding	2.17	Occupational Stress Management Program
2.6	Safety Signs	2.18	Respiratory Protection Program and Equip.
2.7	Hazardous Substance Control	2.19	Blood Borne Pathogens Program
2.8	Welding and Cutting Safety	2.20	Food Safety Program
2.9	Hand Tools	2.21	Lifting Equipment, Gear and Records
2.10	Powered Hand Tools	2.22	Compressed Gas Cylinders: Pressure Vessels
2.11	Lockout, Tag-out and Tryout (Energy Control)	2.23	Motorized Equipment: Checklist, Licensing
2.12	Labelling of Switches, Controllers, Isolators, etc.		

FIGURE 10.1 Section 2 of the Example SMS (23 elements).

The standard should be applied to all work that is performed in restricted areas such as testing, maintenance, repairing, construction, other works, or on any work site where work is performed for the company either by own employees or contractors.

HOLD TAGS

A hold tag is a prominent warning device, which can be attached and securely fastened to an energy isolating device *in conjunction with a lock*. This is to indicate that the energy isolating device and the equipment being controlled may not be operated until the tag and lock are removed. The tag should endure the environmental conditions, should not be spoiled due to humidity, water, or rust, throughout the lockout period. The tag is integrated with lockout devices in preventing any accidental or unauthorized operation, and warns against removing the tag and lock. The tag should indicate who locked the energy source, the reason for isolation, and the date of the isolation. A tag should *never* be used in place of a physical lock.

LABELING OF SWITCHES, CONTROLLERS, ISOLATORS, DISCONNECTS, AND VALVES (S2, E2.12)

To ensure that the correct equipment, circuit, valve, or process is operated (especially in an emergency), all valves, operating devices, switches, isolators, and contact breakers should be labeled or marked. Regular safety inspections should include examination of these labels. In electrical installations, labeling indicates the voltage of the device, where it is fed from, what it feeds, and other relevant information, and also warns of arc flash danger and level of protective equipment required.

A number of fatal accidents have occurred as a result of confusion with switches, isolators, or valves because they were not clearly marked. Identifying and labeling ensure that the proper equipment or machinery is stopped, started, or isolated when required. Furthermore, it assists employees unfamiliar with machinery, or the plant, to readily identify and locate equipment, circuits, switches, isolators, and valves.

Some of the details on labeling that should be included in the standard are as follows:

- Labelling must be permanent and standardized throughout and in English and other language if applicable.
- All fixed equipment must be clearly labeled with the name by which it is commonly known, that is, Fire Pump no. 2, etc.
- Individual units of a number of similar machines must be identified with numbers and/or letters, with one sign for the complete group, for example: Air Conditioners: A/C # 28, and A/C # 29, etc.
- The source of supply must be indicated on the label, so that the circuit can be traced to the isolator or circuit breaker controlling the unit or circuit.

PERSONAL PROTECTIVE EQUIPMENT (PPE) (S2, E2.13)

A structured approach to personal protective clothing and equipment (PPE) is needed in a safety management system. A planned approach begins with assessing the risks of the work and eliminating the risk as far as practical and financially feasible.

Before issuing personal protective equipment (PPE) to employees, the hierarchy of hazard control should be considered and every effort should be made to eliminate the hazard, or to apply substitution. PPE is the last resort in safeguarding. Once all other mitigation efforts have been exhausted, a PPE risk assessment should be carried out to determine who must wear what items of PPE, and what special tasks require PPE. The PPE elements of the safety system include the following:

- Head protection
- Face and eye protection
- Respiratory protection
- Hearing protection
- Protective clothing
- Hand protection
- Foot protection
- Fall restraint and fall protection
- Specialized protective equipment, etc.

Each one of these items should have a written safety system standard and should be included in regular inspections of the usage, storage, and condition of the equipment. The legal requirements around PPE should be regarded as the minimum standard to be achieved and safety system standards should exceed the minimum requirements.

PPE Job/Task	Risk ⇨	🦺	🎧	👁	🥾	😷
Vehicle driver	Risk when offloading vehicle	✓	✗	✗	✗	✓
Rigger	Work at heights, flying particles due to wind, falling objects	✓	✗	✓	✓	✓
Fitter	Work in noise zones also eye injury potential	✗	✓	✗	✗	✓

FIGURE 10.2   An example of a PPE risk matrix.

PPE RISK MATRIX

A PPE risk matrix (Figure 10.2) is now compiled showing the classification of employee, the tasks, and the appropriate PPE for those tasks and classifications. Instruction on the correct wearing of PPE should form part of the induction and ongoing safety awareness training and toolbox talks. Correct use, storage, and condition of PPE should be inspected on a regular basis and critical PPE items, such as fall protection equipment, should be logged in a register for regular inspections.

ACCEPTANCE OF USAGE

Once the PPE risk assessments have been conducted and the correct PPE has been purchased, training in the correct use of the equipment must be provided to all who are involved in the PPE program. Once the training is completed, a written *acceptance of use* document should be signed by employees stating that they understand the need for the wearing of PPE, and that they will abide by the safety rules concerning the use and wearing of the equipment. It must be emphasized that PPE is the last resort to the prevention of injuries in the workplace and that all other means of eliminating the hazard must be applied before the final resort of issuing PPE.

FALL PROTECTION (S2, E2.14)

A risk assessment of the work being done will indicate the need for a fall protection program as part of the safety system. Where a fall hazard is present, and to eliminate injuries caused by falls, a personal fall arrest system should be used. If an employee is exposed to falling 6 ft (2 m), or more from an unprotected side or edge, they must wear a personal fall arrest system to protect themselves. A personal fall arrest system comprises three key components—the anchorage, body wear, and connecting device.

BODY WEAR

Body wear is the personal protective equipment worn by the worker such as a full-body harness, which is the only acceptable body wear for correct fall arrest. The body wear should be selected based on the work to be performed and the work environment. It should be remembered that the side and front D-rings are for positioning only.

CONNECTING DEVICE

The connecting device is the critical link which joins the body wear to the anchorage and anchorage connector such as a shock-absorbing lanyard, fall limiter, self-retracting lifeline, rope grab, etc.

POSITIONING DEVICE SYSTEMS

A positioning device system is a body belt or full body harness, rigged to allow an employee to be supported on an elevated vertical surface, such as a wall, and work with both hands free while leaning.

Positioning device systems are primarily used for safe positioning of employees working at heights and are not fall arrest systems. Safety belts and web lanyards can be used as a positioning device system. This system is to be set up so that workers can free fall no further than 2 ft (0.6 m). They should be secured to an anchorage capable of supporting at least twice the potential impact load of an employee's fall or 3,000 lbs. (1,360 kg) whichever is greater. Requirements for snap-hooks, D-rings, and other connectors used with positioning device systems must meet the same criteria as those for personal fall arrest systems.

TRAVEL RESTRAINT

Travel restraint is used in a situation in which workers on an elevated surface use a body belt or harness with a lanyard, to keep them from being able to reach the fall hazard. Travel restraint is useful in areas where sufficient anchor points for fall arrest are not available or where a fall could cause other problems.

RESPONSIBILITY AND ACCOUNTABILITY

Managers, supervisors, and contractors (Responsible Persons), should be responsible for ensuring that conformity is maintained regarding the organization's fall protection standard. Managers should be made responsible for the fall risk assessment and identification of areas, jobs, and tasks which require the use of personal fall arrest systems. Managers, supervisors, and section heads should be held responsible for ensuring that employees (and contractors in their area) use personal fall arrest equipment when and where required.

Each employee should be held accountable for the wearing of fall protection in required areas, and during specified tasks. All employees who work at heights on a regular basis must undergo an annual occupational medical examination.

INDUSTRIAL (OCCUPATIONAL) HYGIENE ELEMENTS

Industrial hygiene forms a large part of the overall safety management system and has a number of safety system elements which vary from industry to industry.

These elements are normally coordinated by the Industrial Hygienist and are included in Section 2 of the Example SMS.

ERGONOMICS (S2, E2.15)

Since many occupational injuries experienced in the United States are related to ergonomics, this element is of utmost importance in the system. Initial and ongoing ergonomic surveys should be conducted to identify and eliminate sources of ergonomic related injuries, such as repetitive motion disorder, etc. As ergonomics is a specialized science, some organizations appoint a fulltime Ergonomist.

HEARING CONSERVATION (S2, E2.16)

Since noise induced hearing loss is permanent and irreversible, a hearing conservation program is necessary wherever the organization identifies possible noise zones. The elimination of noise zones is the first option and reduction of sound levels is the second. If there is no other way to reduce the exposure of employees to high levels of sound (85 dB or more), a hearing conservation program must be introduced. Employees constantly exposed to noise zones should participate in this program. A hearing conservation program will consist of the following steps:

1. Comprehensive sound level measurements
2. Attempts to reduce noise levels
3. Reduction of employee exposure to noise zones
4. Demarcation and sign posting of noise zones
5. Initial hearing acuity testing and recording
6. Selection, issue, and training in PPE
7. Frequent employee hearing acuity testing (dependent on certain factors)
8. Action taken if hearing loss identified
9. Ongoing monitoring

OCCUPATIONAL STRESS MANAGEMENT PROGRAM (S2, E2.17)

Another element in the safety management system which falls under the Industrial Hygiene elements is occupational stress. Occupational stress is slight temporary physical and/or mental overload caused by overwork, fatigue, depression, worry, and other factors, and could occur during employment. This may be partially caused by external factors such as family life, etc. As a result of tremendous work pressure and maintaining a high standard of living, emotional stress normally builds up within a person. A number of factors contribute to this emotional stress, as individuals are daily in contact with managers, peers, fellow workers, spouses, friends, and children. Interaction with these people could lead to the creation of emotional stress. People lose faith in themselves, they lose confidence and worry. This constant worrying then builds up more stress and leads to a situation where the person is severely stressed.

High-risk Acts

Occupational stress (such as that experienced by air traffic controllers, among others), puts the victim in a position where their full concentration is not always on the job and this in turn causes them to commit high-risk acts or create unsafe conditions, which can lead to accidents. The productivity within an organization is greatly reduced as a result of the stressed person's inability to concentrate on the task at hand and produce efficiently. From a safety point of view, stress could lead to the person taking chances and not following the laid down safety procedures. The safety system should monitor for signs of stress and take the necessary actions before the employee puts him or herself and others at risk.

Heat and Cold Stress

Risk assessments should indicate the heat stress load on employees who work in hot and humid climates and work conditions. In some countries, this element is regarded as their highest risk area and they have comprehensive heat stress management programs, which include supplying employees with ice water, shade provision, rest periods, heat stress monitoring, and acclimatization, and no work periods during the hottest time of the day.

According to the Ohio State University (OSU) *Heat and Cold Stress Program*:

> Working in extreme temperatures (hot or cold) can overwhelm the body's internal temperature control system. When the body is unable to warm or cool itself, heat or cold related stress can result. Heat and cold stress can contribute to adverse health effects which range in severity from discomfort to death.
>
> Environmental Health and Safety (EHS) has developed this Heat and Cold Stress Program to minimize the effects of heat and cold stress on employees. This program contains the procedures and practices for safely working in temperature extremes.
>
> The Occupational Safety and Health Administration (OSHA) does not currently have specific standards for heat or cold stress. However, the Occupational Safety and Health Act of 1970, *General Duty Clause* (Section 5(a)(1)) states that: "*Each employer shall furnish to each of his employees employment, and a place of employment, which are free from recognized hazards that are causing or are likely to cause death or serious physical harm to his employees.*" In addition, 29 CFR Subpart I relating to personal protective equipment requires employers to provide protection to employees exposed to hazards in the workplace. The OSHA website contains Fact Sheets and Guidance Documents that relate to heat and cold stress that have been incorporated into this program. (Ohio State University, website) (p. 3)

ILLUMINATION (S5, E5.7)

Illumination of the work areas is critical to safety and the purpose of this element of the safety management system is to ensure that both natural and artificial lighting levels are within prescribed standards. This element is discussed in detail in Section 5 (Element 5.7) of the Example SMS.

VENTILATION AND AIR QUALITY (S5, E5.8)

This element of the safety system falls under the workplace environment section of the Example SMS, Section 5, and cross references to the industrial hygiene elements. The purpose of the element which is discussed more fully under Section 5 (Element 5.8) is to establish controls for the safe and effective use of natural and artificial ventilation sources, ensure that ventilation systems are specifically matched to the workplace hazards, and establish performance, maintenance, and repair criteria for ventilation systems.

RESPIRATORY PROTECTION PROGRAM AND EQUIPMENT (S2, E2.18)

The control of respiratory hazards shall be accomplished as far as feasible through the implementation of accepted engineering control measures, and whenever engineering controls and/or substitution are deemed not to be feasible, or while they are being implemented, appropriate respiratory protection should be provided and used.

The safety system standard should establish controls for the safe and effective use of respiratory protection devices for potentially hazardous occupations and tasks, and ensure that employees do not experience occupationally related respiratory damage and/or disease. The company should establish medical surveillance requirements for the use of respiratory protection devices even if not required by safety regulations. Procurement standards for respiratory protection devices should form part of this standard.

RESPIRATORY PROTECTION PROGRAM

The organization should administer a continuing, effective respiratory protection program, and respirators shall be provided when such equipment is necessary to protect the health of the employee, as determined by air quality risk assessments. Employees who meet the following criteria should participate in the organization's respiratory protection program:

- Employees who wear or may wear a respirator
- Employees who supervise personnel who wear respirators
- Emergency response personnel
- Wearers of SCBA sets (Self Contained Breathing Apparatus)

Job Safe Practices (JSPs)

JSPs should address respiratory hazards for each operation where the use of respiratory protection equipment is or may be required, as well as respirator use for each location where an emergency response to a respiratory hazard may exist.

BLOODBORNE PATHOGENS PROGRAM (S2, E2.19)

Depending on the nature of the industry, a safety system element and program on bloodborne pathogens may be needed. According to OSHA:

Bloodborne pathogens are infectious microorganisms in human blood that can cause disease in humans. These pathogens include, but are not limited to, hepatitis B (HBV), hepatitis C (HCV) and human immunodeficiency virus (HIV). Needlesticks and other sharps-related injuries may expose workers to bloodborne pathogens. Workers in many occupations, including first responders, housekeeping personnel in some industries, nurses and other healthcare personnel, all may be at risk for exposure to bloodborne pathogens.

What can be done to control exposure to bloodborne pathogens?

In order to reduce or eliminate the hazards of occupational exposure to bloodborne pathogens, an employer must implement an exposure control plan for the worksite with details on employee protection measures. The plan must also describe how an employer will use engineering and work practice controls, personal protective clothing and equipment, employee training, medical surveillance, hepatitis B vaccinations, and other provisions as required by OSHA's *Bloodborne Pathogens Standard* (29 CFR 1910.1030). Engineering controls are the primary means of eliminating or minimizing employee exposure and include the use of safer medical devices, such as needleless devices, shielded needle devices, and plastic capillary tubes. (OSHA website, 2015d.)

FOOD SAFETY PROGRAM (S2, E2.20)

If refreshments or meals and snacks are prepared and supplied to employees, or sold to the public, a food safety plan should form part of the safety management system. Dependent on the risk, type, and amount of the food stored, prepared, sold, or served, the plan should cover the receiving, storage, defrosting, preparing, and selling, or serving of the food.

A food safety plan (FSP), also often referred to as a HACCP Plan (Hazard Analysis Critical Control Point), is a written procedure that will help eliminate, prevent, or reduce food safety hazards that may cause customers or employees to become ill or injured (Figure 10.3). Food safety plans begin at the receiving and storage stage where the food enters the premises until the point where it is served or purchased.

It is desirable for every operator of a food service establishment and food premises, where carcasses are handled or where food is processed or prepared, to develop, maintain, and follow a food safety plan to ensure that a health hazard does not occur in the operation of the facility. A food safety plan must be completed and managed by qualified and approved personnel.

Some of the most common practices that lead to foodborne illnesses include: improper cooling and cold storage, advanced preparation, inadequate reheating, cross-contamination, and inadequate cooking. A food safety plan focuses on the

Preparation step	Critical control points	Potential hazards	Critical limits (food safety standards)	Monitoring actions	Corrective actions
Receiving	Yes	Contamination growth of pathogens	Food is obtained from approved sources Refrigerate food at 4°C or less Food is wholesome, free from pests, packaging secure	Verify with supplier if in doubt Check temperature and record Visual inspection	Return food to supplier
Storage	Yes	Growth of pathogens	Store at 4°C or less Frozen food at −18°C or less Thaw frozen food: In cooler or refrigerator/ under cold water/in microwave, prior to use	Check temperature and record Check temperature Observer thawing process	Adjust temperature settings Move food to alternative unit Discard food held above 4°C for more than 2 hours Modify practices, discard contaminated food

FIGURE 10.3   An extract from a food safety plan.

critical steps within the preparation of the food to prevent these practices from occurring. The food safety plan should be tailored to match the process followed by the organization.

HYGIENE AMENITIES (S5, E5.9)

Facilities provided for employees such as restrooms, lunchrooms, emergency showers, and eyewashes, change rooms, and showers must be adequate and maintained in a clean, hygienic state, and the coordination of this objective is usually done in conjunction with the Industrial Hygiene department. This element is discussed in detail in Section 5 (Element 5.9) of the Example SMS.

LIFTING EQUIPMENT, GEAR, AND RECORDS (S2, E2.21)

Lifting gear and equipment has greater potential for causing damage or personal injury if such equipment is incorrectly used, not maintained, or operated by untrained, unauthorized personnel. Regular inspections of equipment, training, and authorization of personnel are essential to ensure that equipment is in good working condition and that it is correctly used.

Lifting Machines are equipment used for raising, lowering, pushing, or pulling a load and include the following:

- Overhead and mobile cranes
- Crawler locomotive and truck cranes
- Electric, air and hydraulic hoists and winches, jib cranes, mechanical and hydraulic jacks

- Man-lifts and fork-lift trucks
- "Cherry Picker" type man hoists or similar equipment
- Chain and wire rope blocks. A block consists of one or more sheaves (pulleys) within a holder or housing

Lifting tackle is equipment used to provide a connection between the lifting machine and load such as: chains, wire ropes, fiber rope slings, synthetic webbing, hooks, shackles, and swivels, etc. These items should be individually identified and placed on a regular inspection system.

MAJOR LIFTS

Major lifts of heavy and awkward loads, or unusual, non-routine lifts should be subjected to a risk assessment before the work commences, and a safety plan for the lift compiled and followed. All lifting machines and tackle must be of good construction and comply with OSHA 1910.179 and 1910.184, as an example.

RESPONSIBILITY AND ACCOUNTABILITY

Managers, supervisors, and contractors (Responsible Persons) should be responsible for maintaining a list of persons authorized to operate lifting machinery in their respective areas. Overhead and mobile crane operators, and other nominated persons are to be made responsible for performing a daily inspection of all functional operating mechanisms, air and hydraulic systems, chains, rope slings, hooks, and other lifting equipment. Daily check sheets must be completed and a checklist file should be maintained by Responsible Persons.

OPERATOR TRAINING

Managers, supervisors, and contractors should be made responsible for ensuring that designated personnel are trained in the use, maintenance, and inspection of lifting equipment, and that employees are educated in standard hand signals. There should only be one set of hand signs used and since different countries use different signs, agreement to which signals will be used must be agreed to before the lift. Internal operator licenses should be issued to all lifting equipment operators once they have attended the required training. These could be renewed annually. All crane operators should attend and pass an annual medical examination.

INSPECTIONS

Lifting tackle must be inspected before each use and the user is responsible for these inspections. Managers, supervisors, and contractors are responsible for ensuring that suitably trained persons carry out inspections of lifting equipment at scheduled intervals and that the results are recorded on the relevant checklist. Unsafe equipment should not be used until repaired.

COMPRESSED GAS CYLINDERS: PRESSURE VESSELS AND RECORDS (S2, E2.22)

An example of a standard for gas cylinders should contain at least the following basic rules:

- Only authorized and trained personnel are permitted to use welding, cutting, lancing, or brazing equipment.
- Compressed gas cylinders are to be regularly examined for obvious signs of defects, deep rusting, or leakage.
- Care must be used when handling and storing cylinders, safety valves, and relief valves to prevent damage.
- Precautions must be taken to prevent the mixture of air or oxygen with flammable gases, except at a burner or in a standard torch.
- Anti-flashback devices should be fitted to all oxyacetylene cutting and welding sets.
- Only approved apparatus (torches, regulators, pressure reducing valves, acetylene generators, and manifolds) should be used.
- Cylinders are to be kept away from sources of heat and away from elevators, stairs, or gangways.
- Do not use cylinders as rollers or supports.
- Empty cylinders must be appropriately marked and their valves kept closed.
- All cylinders are to be kept in storage racks and are to be securely chained to prevent them from falling.

VESSELS UNDER PRESSURE

Other vessels under pressure should conform to the safety regulations and must be tested and inspected according to manufacturer's specifications or guidelines of an approved authority. These would include air receiver tanks, boilers, and other vessels under pressure. Records should be kept of all inspections, tests, and repairs.

MOTORIZED EQUIPMENT: CHECKLIST, LICENSING (S2, E2.23)

A high risk within any organization is the risk of motorized vehicles such as cars, trucks, mobile cranes, and fork-lift trucks, etc. A comprehensive vehicle safety program should be in place as part of the safety management system. Being one of the most important elements in the system, a detailed standard should encompass the requirements of users, drivers, operators, and the inspectors and maintainers of the equipment. The program should incorporate the selection of operators, training, licensing, pre-start inspections, operating rules, and other pertinent information.

Daily and pre-start checks should be documented, and drivers of equipment should be clearly identified. Seat belt usage should be made mandatory for all vehicles equipped with seat belts. A simple guideline is "If there is a seat belt, it shall be worn." Traffic rule obedience should be enforced and vehicles that use public roads could have vehicle behavior feedback messages such as "How is my driving?" with a contact phone number displayed. This simple system will give management an idea as to how well company vehicle drivers are adhering to the rules of the road.

11 Emergency Preparedness, Fire Prevention, and Protection

EXAMPLE SMS: SECTION 3

This section of the Example Safety Management System (Example SMS) contains 12 main elements covering the basics of fire prevention and emergency preparedness (Figure 11.1). A comprehensive safety system would also have many sub-elements for each element.

WORLD'S BEST PRACTICE

One of the best standards available for fire prevention and protection standards, and information are the NFPA (National Fire Prevention Association) Standards, the NFPA 101: *Life Safety Code.*
Document Scope of NFPA 101

> The *Life Safety Code* is the most widely used source for strategies to protect people based on building construction, protection, and occupancy features that minimize the effects of fire and related hazards. Unique in the field, it is the only document that covers life safety in both new and existing structures. (NFPA website, 2015)

Any organization wishing to maintain a best-in-practice safety system should ensure that the requirements and recommendations of this code are incorporated into the system's standards and practices.

FIRE PREVENTION AND PROTECTION COORDINATOR (SECTION 3, ELEMENT 3.1) (S3, E3.1)

A person with knowledge and practical experience in firefighting and fire prevention tasks, and emergency response, should be appointed for each site within the company, to coordinate and control the company's overall fire prevention and emergency response initiative. This appointment is normally supplementary to the appointee's scope of employment and job description, but could be a full-time appointment in some cases, and should be in writing with basic duties and functions detailed.

The appointee must have adequate training, could be a member of the full-time fire department if available on the site, and will be responsible for fire prevention tasks and the emergency response program.

Number	Element Title	Number	Element Title
3.1	Fire Prevention and Protection Coordinator	3.7	Alarm System
3.2	Fire Risk Assessment	3.8	Fire-Fighting Drill and Instruction
3.3	Fire Extinguishing Equipment	3.9	Emergency Planning
3.4	Locations Marked, Floor Clear	3.10	First-aid, Emergency Responder and Facilities
3.5	Maintenance of Equipment	3.11	First-aid (First Responder) Training
3.6	Storage of Flammable and Explosive Material	3.12	Security System

FIGURE 11.1   Section 3 of the Example SMS.

FIRE RISK ASSESSMENT (S3, E3.2)

The next element in this section is the fire risk assessment. The fire risks of the organization should be identified through fire risk surveys performed by company appointed fire officials, insurance surveyors, or both.

The fire risk survey should:

- Describe the fire hazard.
- Rank the hazard priority (High, Medium, Low).
- Give details of the precautionary measures taken.
- Describe the procedure to be followed in case of a fire occurring.
- Consider the safety, health, and environmental risks associated with fires.
- Include hot work permitting system, etc.

Fire risk assessments and the action to be taken should be available for review and audit. The responsible group (company fire officials and/or insurance surveyors) should formulate appropriate plans to manage the fire risks of the organization. The plan to reduce and manage the fire risks of the organization should be based on the formal fire risk assessment survey.

The local fire services department could be invited to visit the premises and be requested to issue a report stating that all its requirements have been met with regard to fire prevention procedures, equipment, and escape routes. Ongoing liaison should be maintained, and records of these visits and consultations kept.

FIRE EQUIPMENT

The company appointed Fire Coordinator should ensure that the correct types of fire equipment have been provided for each identified risk area. This includes vehicles and equipment. All fire equipment should be strategically located in relation to the fire risk, for example, at the exit to an office or room, not inside the room. The coordinator should maintain a layout plan of the premises on which all fire equipment and critical valves are indicated.

The emergency plans will be reviewed by the Fire Coordinator to ensure that the necessary fire prevention and protection measures are in place. The measures should include regular fire risks inspections. The inspection should ensure that no extinguishers are located too close to the fire hazard, thus rendering them inaccessible during a fire situation, and should also note other deviations from accepted practice. The monthly inspection checklist completed by Safety Representatives could include checking that fire equipment is available and is in a good state of repair.

FIRE EXTINGUISHING EQUIPMENT (S3, E3.3)

It is essential that the fire risks of the organization are correctly identified and assessed to ensure that the appropriate fire detection, protection, and extinguishing equipment, are provided at the appropriate locations. This includes fixed fire systems, passive protection, as well as firefighting and rescue equipment.

The provided equipment should be strategically placed, well signposted, and easily accessible at all times. Personnel should be trained in the use of firefighting equipment, and must be shown how to operate fire alarms when necessary. Specialized fire systems should be installed and maintained by specialists, according to the NFPA recommended codes.

LOCATIONS MARKED, FLOOR CLEAR (S3, E3.4)

To clearly identify the positions of firefighting equipment and systems, and to ensure that they are readily accessible, "keep clear" areas should be demarcated and maintained below their position. Mounted firefighting equipment backgrounds are to be erected where practical. All demarcation, signs, and notices should be kept clear and visible. Locations of equipment should ideally be marked and numbered. All firefighting equipment should also be numbered, and equipment and systems should be accessible and not obstructed.

Locations of fire equipment should be marked with appropriate signage, and should be posted above the equipment. Arrows in the form of safety signs can be used to indicate fire extinguishers situated around corners, or where not readily visible. Arrow signs should be located for maximum visibility.

MAINTENANCE OF EQUIPMENT (S3, E3.5)

Whenever fire equipment is provided to protect a particular risk, it is important that the integrity of the equipment be maintained. This includes ensuring that portable equipment is in good order and accessible, and that fixed installations are maintained in operational order.

The organization should appoint suitably qualified persons to inspect all firefighting equipment at regular intervals. All equipment should be inspected, as a minimum, monthly by a suitably trained and appointed person who must be competent to carry out these inspections.

INSPECTIONS

No equipment should be excluded from the inspection, and must include at least the following: fire extinguishers, hose reels, fire hydrants, fire trolleys, fire vehicles, deluge systems, sprinkler installations, etc. Fixed firefighting equipment should be serviced at least annually, or more often if the specific prevailing conditions on the site require that it be serviced more often (e.g., highly corrosive atmospheres), or as determined by international standards.

STANDBY EQUIPMENT

Standby units of the correct type should be available in place of units that have to be serviced offsite, or the service company should provide standby units in the place of any units which might be removed from the site for servicing. Whenever a fire extinguisher is discharged, immediate action is to be initiated to replace or refill the extinguisher. Pressure tests should be conducted on fire extinguishers according to legal or international fire standards. All sprinkler installations should be inspected once a year by an approved sprinkler inspection authority.

PRESSURE TESTS

Records are to be kept of pressure tests performed on fire equipment pressure vessels. Proof of tests should be maintained when outside contractors are used for maintenance. All automatic fire suppression systems should be checked and maintained in accordance with established standards, and at prescribed intervals. Booster pumps should be run and tested as per prescribed standards. Pipelines should be inspected for leaks, and repaired as necessary.

STORAGE OF FLAMMABLE AND EXPLOSIVE MATERIAL (S3, E3.6)

Flammable and explosive materials do not only pose a fire hazard, but also pose a hazard to facilities and personnel in proximity. There is an inherent hazard in these materials which need to be considered in the planning for storage and use.

The primary method of reducing the risk posed by flammables, is to restrict the quantities to acceptable levels during storage and use. A flammable materials risk assessment will determine if the quantity of flammable liquids kept on a site warrants the building of one or more flammable liquid stores.

FLAMMABLE LIQUID STORES

Where necessary, formal approval for a flammable store may be required from the relevant authority. A copy of the approval certificate, if applicable, should be displayed at the store. The maximum volume of flammable substance permitted should be posted on the outside or entrance door.

The Fire Coordinator or other qualified personnel should consider the chemical compatibility of the various chemicals and provide advice on segregation where necessary. Each department using flammable and explosive materials should appoint a competent person (supervisor or line manager) to take responsibility for the storage of all flammable material in their area.

Storage

The storage and issue of chemicals must comply with legislation, international and company standards and procedures. Containment of flammable liquids shall be determined at 110% of the capacity of the store. All electrical equipment used within the unsafe zone of a flammable substance needs to be intrinsically safe. Intrinsically safe lighting should be placed on planned maintenance and be inspected as to the integrity of the seals on an annual basis.

Labeling and Signs

All items in flammable stores must be properly and clearly labeled as to the content and volume, and be arranged neatly on racks and shelves. No combustible materials such as wood, rags, carton boxes, etc., should be kept in a flammable liquid store or cabinet.

"No open flames" and "No smoking" safety signs shall be displayed in the vicinity of the flammable liquid store or, where applicable, on the doors of the flammable liquid cabinets. The use of cell phones, and radios, and any similar potential source of spark should be prohibited in these areas.

Flammable Liquid Cabinets

Where there is no need for a flammable liquid store, all flammable liquids should be stored in a metal cabinet with doors and gauzed ventilation holes, as approved by the Fire Coordinator.

Usage

Flammable liquids shall be issued only on a need-to-use volume basis and strict control is exercised to ensure that persons do not draw more than what is needed for the specific job. Bonding cables or chains should be available to connect to (bond) containers of highly volatile liquids when decanting takes place, and drip trays should be provided. Drip trays should be emptied after decanting so as to ensure that a hazardous atmosphere is not created by evaporating liquids from drip trays.

Spill absorbent material should be available where there is a risk of spills. Sawdust or rags should not be used to absorb flammable liquids, and sufficient numbers of the correct firefighting equipment should be made available in close vicinity of the flammable liquid store.

FIRST AID MEASURES

Material Safety Data Sheets (MSDS) information, including handling, storage, emergency, and first aid instructions should be available at areas of bulk storage, and all employees using flammable materials shall be trained and informed about the hazards and emergency procedures related to the chemicals they are using. If an organization utilizes explosives, special storage magazines, which should conform to regulatory requirements, are required.

ALARM SYSTEMS (S3, E3.7)

The effective communication of imminent danger to all parties on site during emergency situations will facilitate appropriate response and ensure that exposure to danger is limited to those dealing with the emergency situation.

The objective of the requirements of this safety system element should be to ensure that all alarm systems are in working order, and can effectively be activated in an emergency situation, and to ensure that all persons on site are familiar with the alarms.

REQUIREMENTS

A fire alarm with a backup in case of a power failure should be installed on all the organization's premises. The backup alarm may be an uninterrupted power supply (UPS) system, a hand operated bell, siren, or a portable compressed air operated aerosol horn. The Fire Coordinator should ensure that all the fire alarm activating points are indicated on a layout plan of the premises, and erected at critical positions. Safety staff should verify that all employees and contractors in their area or department know the location of the fire alarm activation point, and are familiar with the sound of the fire alarm, through checks during inspections.

The Fire Coordinator should test the fixed fire alarm at least once every three months and keep a record of these tests which should be made available for inspection. In remote locations where portable alarms are used, the department should inspect and test the manual alarm once every three months.

Fixed alarm systems must be included on planned maintenance. The instructor delivering the safety induction training must ensure that all new employees are notified of the sound of the different alarms during their training. Any isolation of an alarm system due to maintenance or malfunction should be backed up by a manual alarm system and the persons in the area should be informed.

FIRE FIGHTING DRILL AND INSTRUCTION (S3, E3.8)

FIRE TEAMS

Dependent on the fire risk assessment, an organization may have to form fire teams made up of employees. These team members need to be trained in the basics of firefighting and need to practice on a regular basis. The appointed Fire Coordinator should command the fire teams and supervise their exercises and practical training. Teams of first responders, firefighters, as well as technical staff, may be required by some entities. Contractor sites should follow the requirements of the organization and also fulfill the requirements of the standard. Mines have specially trained and equipped rescue teams available on all shifts, in case an emergency rescue needs to be carried out. Their requirements are normally prescribed by safety regulations.

FIRE DRILLS

A six-monthly dry run would be a minimum requirement for fire drills, and these could coincide with the regular evacuation drills. In some cases, a scenario is planned, and some employees pose as injured persons to test the reaction of the first responders and the emergency plan. The intensity and frequency of these drills vary according to the nature of the business.

EMERGENCY EVACUATIONS

Every building and facility within the company operation areas should have its own documented emergency evacuation plan and should conduct an emergency evacuation for their respective buildings as per their plan. This may be done simultaneously with a fire drill.

The objective of the emergency evacuation plan is to:

- Provide an orderly evacuation plan for all employees.
- Ensure all exit routes, doors, and emergency staircases are not obstructed and can be used in an orderly way during emergencies.
- Familiarize building occupants with the means of escape routes from the building.
- Ensure fast, organized and smooth evacuation of buildings during emergencies.
- Test the working conditions and effectiveness of all fire and emergency equipment and systems for all buildings.
- Identify any weakness in the evacuation plan.

The evacuation drill must be conducted at least twice a year for each occupied building which includes main sites, general offices, warehouses, workshops, and other workplaces.

EMERGENCY PLANNING (S3, E3.9)

No matter how good or efficient an organization's safety management system, the possibility still exists for the company to experience a natural or man-made emergency situation. It is through good planning and experience that the severity of the situation can be limited and the effects of the situation mitigated.

RESPONSIBILITY AND ACCOUNTABILITY

Each site manager, supervisor, and contractor should ensure a written detailed emergency plan is available for their respective facility or building. The emergency plan should address all types of expected events. It should include actions to be taken before, during, and after the event, and the roles of all participants. They should ensure the periodic testing of the emergency plan. All likely scenarios should be practiced.

The persons responsible should advise executive management on the appropriate systems and facilities required for effective emergency response, and the executive management should provide the necessary resources and infrastructure for this emergency response.

EMERGENCY PLAN

The first step in compiling an emergency plan is to identify the credible emergency scenarios that may occur at the organization such as the following:

- Fire
- Gas fire or explosion
- Fuel or chemical leakage
- Natural disasters (floods, volcanic eruptions, earthquakes, tsunamis, and other geologic processes)
- Nuclear radiation
- Rescue of personnel
- Business continuity, inter alia

The appropriate managers should ensure that the credible scenarios are covered in the emergency plans. The Fire Coordinator should ensure that the fire team is trained in emergency situations and that they participate in emergency drills. The organization should liaise with the local fire authority regarding assistance to supplement the on-site fire teams where the situation is beyond their capabilities or control. The environmental department should utilize the environmental aspect register and significance rating process to identify environmental emergency scenarios, and notify the coordinator for inclusion in the team drills and training.

INCIDENT COMMANDER

Top management should appoint a responsible person to act as the incident commander who should coordinate all emergency procedures. The Incident Commander should use the emergency plan to anticipate likely scenarios and develop tactical operations. He or she should obtain and post floor plans with evacuation routes and other information as specified in the emergency plan, schedule fire drills and emergency evacuations, and supervise and determine the roles of outside service entities participating in the drills. During exercises, the Incident Commander should wear a reflective vest suitably labeled for easy identification.

EMERGENCY INSTRUCTIONS

An up to date emergency call chart for external assistance, which lists all the appropriate contact numbers, should be maintained. This chart should be displayed at areas where the external assistance is likely to be summonsed, and will include control rooms, emergency control center, the security gate office, and the safety or fire department.

Evacuation routes should be clearly signposted and illuminated for night workers. Evacuation procedures should be prominently displayed in all areas, and assembly points must be situated away from any danger zone and clearly signposted.

FIRST AID, EMERGENCY RESPONDER, AND FACILITIES (S3, E3.10)

First aid facilities should be provided throughout the workplace to deal with first aid cases. These facilities include the provision of first aid boxes, stretchers, eye washes, etc. Some enterprises have on-site first aid rooms attended by trained first responders. The provision of adequate facilities would be dictated by the size, nature, and location of the organization. Each organization should undertake a survey of what first aid response they require, and provide and maintain the facilities accordingly. The emergency plan would also help determine the facilities needed. The legal requirements should be seen as a minimum requirement.

FIRST AID (FIRST RESPONDER) TRAINING (S3, E3.11)

The training of employees in the basics of first aid is a requirement of a good safety system, as is also required by safety regulations in many instances. Risk assessments will indicate the need for this training. Not only does this training make employees more aware of safety, but it also provides trained personnel in case of accidental injury or medical emergency. An undertaking should train a minimum of 5% of its workforce in first aid, annually. Each first aider should also be identified, and their name and location posted on the safety notice board.

SECURITY SYSTEM (S3, E3.12)

Since security personnel are normally on duty 24 hours a day, they can report unsafe situations noted during their routine patrols for rectification by maintenance crews. Street lamps and other lights that are not working can be reported by them, as well as structural damage that might have been caused by vehicles or other devices.

In many instances, they are trained fire fighters and form part of the fire team. During their patrols, they could also ensure that no fire equipment has been tampered with or is missing.

A good security system will prevent unauthorized entry to the workplace, thus helping secure the safety and wellbeing of employees. They could also intervene in work related aggression between employees and prevent workplace violence injuries. Vehicles entering or leaving the plant can be checked by security checkpoints to ensure that they are correctly loaded, have all lights functioning correctly, and that the driver and co-driver wear their seat belts. Security officers could also participate in incident recall sessions.

12 Accident and Near Miss Incident Recording and Investigation

EXAMPLE SMS: SECTION 4

Section 4 of the Example Safety Management System (Example SMS) is Accident and Near Miss Incident Recording and Investigation, and consists of nine elements (Figure 12.1).

ORGANIZATION FAILURES

Accidents, in simple terms, are symptoms of a failure of the organization to adequately identify and mitigate the risk. Traditionally, accidents were investigated to ascertain who it was who caused the event, and once that was established, the investigation ended without identifying or rectifying the real problem. A structured safety management system (SMS) views an accident, or near miss incident, as an opportunity to identify and fix weaknesses in the organization's approach to safety. A modern approach to investigating accidents is finding the facts and not the faults.

CONFUSION

Often confusion exists in the safety and health field concerning the terms, *accident*, *incident*, and *near miss*. In the past, the term *incident* was used to describe *near misses*, but since the modern approach is to term *accidents* (loss producing undesired events), *and near miss* (non-loss causing events), incidents, confusion still exists between what is an *accident*, *incident*, and *near miss*.

Often, the word *accident* is also replaced by the term *incident*, which leads to confusion, therefore this publication has referred to near misses, close calls, and near hits as *near miss incidents* to remove any confusion concerning terminology.

DEFINITIONS OF AN ACCIDENT

Here are some definitions of an accident to indicate that there is general consensus that an event termed an *accident* results in some form of loss, either to an individual, property, organization, or all of these.

- An accident is an undesired event often caused by unsafe acts and/or unsafe conditions, and results in physical harm to persons, and/or damage to property and/or business interruption.

171

Number	Element Title	Number	Element Title
4.1	Occupation Injury and Disease Records and Registers	4.6	Loss Statistics Kept
4.2	Internal Accident and Near Miss Incident Reporting and Recording	4.7	Cost of Risk: Apportioning of Costs
4.3	Internal Accident and Near Miss Incident Investigation	4.8	Near Miss Incident and Accident Recall
4.4	Near Miss Incident Reporting System	4.9	Return to Work Program
4.5	Occupational Injury, Disease, Damage Statistics		

FIGURE 12.1 Section 4 of the Example SMS.

- An accident is the culmination of a series of activities, conditions, and situations which ends in injury, damage, or interruption.
- An accident is the occurrence of a sequence of events that usually produces unintended injury or illness, or death, and/or property damage.
- An accident is an undesired event, or sequence of events causing injury, ill-health, and/or property damage.
- An accident is an undesired event, which results in harm to people, damage to property and/or loss to process.

The *contact with a source of energy* phase in the accident sequence is traditionally referred to as the *accident*, which is incorrect, as the accident is the total sequence of events and the loss (injury and damage), is the last phase, or consequence segment, of the event.

NEAR MISS INCIDENTS

Near miss incidents are near miss events that come close to causing some form of loss as there was an actual flow of, or exchange of energy below the threshold level. In some instances, the flow of energy may have dissipated without making any contact, thus causing no loss. In most cases, the energy does not contact anything, thus causing no harm. In some cases, the exchange of energy was insufficient to cause loss or injury, but the fact that there was an exchange of energy is reason enough to heed the warning. Remember, it is not what happened . . . but what *could* have happened! Near miss incidents are accidents *waiting in the shadows*.

DEFINING A NEAR MISS INCIDENT

Near miss incidents are also known as; *near miss, close call, incident, close shaves, warnings, or near hits*. In the case of moving objects, these events are known as *near collisions*.

A near miss incident is:

- An undesired event, which under slightly different circumstances, could have resulted in harm to people, and/or property damage, and/or business disruption.

- An event that narrowly missed causing injury or damage.
- An incident where, given a slight shift in time or distance, injury, ill-health, or damage easily could have occurred, but did not this time around.

OCCUPATIONAL INJURY AND DISEASE RECORDING (SECTION 4, ELEMENT 4.1) (S4, E4.1)

INJURY AND DISEASE CLASSIFICATION

The objective of this element of the safety system is to ensure that all occupational injuries and diseases are classified according to standard rules and criteria, and that reportable, recordable, and compensatory injuries and diseases are clearly defined. Injury rates cannot be calculated and compared with other organizations, or other similar classes of industry, if the same injury and illness classification system is not used. These classifications are often spelt out in safety regulations, or by national standards, such as ANSI-Z16 *Recording and Measuring Injury Experience* and OSHA (29 CFR 1904). The organization should establish which standard to follow and incorporate those criteria into the elements of the safety system.

RESPONSIBILITY AND ACCOUNTABILITY

Managers, supervisors, and contractors (Responsible Persons), should be held accountable for adhering to this classification and reporting criteria, and for the correct classification of injuries and diseases occurring within their areas of responsibility. Safety staff are to guide and assist managers, supervisors, and contractors in injury reporting and classification activities, to ensure conformity with the organization's standard. Employees should be informed and made responsible to report all injuries and all occupational diseases and illnesses to their immediate superior.

CLASSIFICATION GUIDE

The following is a guideline to Responsible Persons as to what shall be classified as a work related lost time injury, illness, or disease.

- A *work injury* is any injury suffered by a person which arises out of and in the course of their employment. (Wherever the word "injury" is used, it shall be construed to also include occupational disease and work connected disability.)
- An *occupational disease* is a disease caused by environmental factors, the exposure to which is peculiar to a particular process, trade, or occupation, and to which an employee is not ordinarily subjected or exposed outside of or away from such employment.

Local by-laws also prescribe certain circumstances which call for the classification of injuries as work related and these should receive preference over the organization's classification.

Injury and Disease Records or Registers

Irrespective of legal requirements, the organization should keep a record of all minor, lost time, disabling and other injuries and illnesses. This record should give the basic description of the event, the time and date, and the outcomes, and is seen as a requirement over and above the minimum requirements of local safety regulations.

INTERNAL ACCIDENT AND NEAR MISS INCIDENT REPORTING AND RECORDING (S4, E4.2)

A formal, structured system of reporting and recording accidents and near miss incidents should be introduced. Standard forms, either hard copy or online, should be drawn up and made available for reporting purposes.

Internal Accident Reporting

Many companies have a one-page *accident initial report* form which is completed and circulated immediately upon occurrence of an undesired event. This informs all as to what has occurred and provides information and eliminates speculation and rumor concerning the event. This is followed up with the results of the final accident investigation.

A record should be kept of all accidents and near miss incidents. Each event should receive a tracking number so that it can be followed through the system and for later reference.

The organization's standards should call for employees to report all injuries and diseases to their supervisor without repercussion, in case the wound later becomes infected, or minor strain experienced manifests into a more serious injury. The reporting system should create a comfort zone whereby employees feel confident to report all injuries and possible injuries, without fear of reprisal. Reporting procedures should be established and employees notified of these procedures.

INTERNAL ACCIDENT AND NEAR MISS INCIDENT INVESTIGATION (S4, E4.3)

The purpose of both accident and near miss incident investigation is to carry out an investigation into the undesired event, to determine what happened and what can be done to prevent a similar event recurring. If positive preventative measures are not taken after an accident, or high potential near miss incident, the probability of recurrence is great. The investigation procedure would identify the root causes of the event and help determine what steps to take to prevent a similar accident or undesired event.

Post-contact versus Precontact

Accident investigation is a post-contact control activity, as a loss must first occur before the accident can be investigated, and preventative steps can be taken.

Although a post-contact control activity, accident investigation leads to the taking of preventative measures, which in turn is a precontact activity. Near miss incident investigation, on the other hand, is a precontact activity as no loss occurred, but the potential for loss was recognized, and by investigating and taking preventative measures, a recurrence which may have devastating consequence is prevented.

BENEFITS OF ACCIDENT AND NEAR MISS INCIDENT INVESTIGATION

Accident and high potential near miss incident investigations, properly conducted, help identify and quantify the losses incurred by undesired events. They help determine the facts of the situation and also define which accident type occurred. Investigation into the event helps determine where the safety management system failed. Property damage accidents should be viewed as important as injury events, as accidental property damage could have resulted in injury under slightly different circumstances.

The investigation also helps determine the exact extent of the personal injury or disease, the extent of damage, as well as what actually happened. Only by investigating the event, can the physical and occupational hygiene agencies and the agency parts be identified and classified.

Investigation also indicates who was involved at the time of the accident and most importantly, is a means to identify the immediate causes in the form of high-risk acts and unsafe conditions which, once determined, enable the root causes in the form of *personal factors* (both cognitive and non-cognitive) and the *job factors*, to be established and rectified.

Personal Factors

Personal factors could be factors such as stress, lack of skill, lack of knowledge, inadequate motivation, physical or mental shortcomings, or a poor attitude towards safety. The employee could be aware that they were taking a risk (cognitive) or could be unaware that the situation posed a risk (non-cognitive).

Job or Environmental Factors

Job factors could include items such as excessive wear and tear, inadequate safety standards, tools and equipment which are insufficient, poor or no maintenance, no standards for purchasing, lack of maintenance, or poor supervision.

MULTIPLE CAUSES

Very rarely does an accident occur as a result of a single cause. Experience has shown that multiple causes are almost always present in most accidents and near miss incidents. Accident investigation helps determine which causes contributed to the loss, and what steps to take to prevent a recurrence.

INVESTIGATION STANDARD AND FORM

The organization's standard covering accident and high potential near miss incident investigation should include the following:

- The training and development required for internal accident investigators
- A comprehensive accident investigation form that provides a step-by-step methodology of investigating the event
- Witness interview forms
- Immediate and root causes, to allow for immediate cause analysis and root cause analysis to take place
- An action plan to remedy the causation factors allocating responsibility for the actions
- Deadline dates
- A conclusion with action steps that will fix the problem

NEAR MISS INCIDENT REPORTING SYSTEM (S4, E4.4)

Since numerous near miss incidents occur for each serious injury experienced, they should be regarded as warning signs, or accident precursors, and be reported and risk ranked. Those with high potential should be investigated as thoroughly as a loss producing accident. Employees should be made aware of the importance of near miss incident reporting and should be encouraged and rewarded for reporting these occurrences. The greatest reward to a reporter of a near miss incident is to see some action being taken on their report. The system makes employees feel a part of the safety system and they realize that they can play an important role, while not fearing retribution for reporting. This is why a near miss incident reporting system must provide a safe space to allow for anonymous reporting, with no questions asked.

BOOKLETS

Pocket-size forms or booklets should be available throughout the plant, so that anyone can fill in a form to report an event. Boxes for the deposit of the completed forms should be erected throughout the plant as well, so that near miss incident reports can be collected and processed. The form could include a mini risk matrix on which the reporter can risk rank the event. It should include a space where they can write a suggestion on how to rectify the situation, so as to prevent a similar event from occurring.

Some organizations also use a safety telephone hotline for reporting of any safety issue, including near miss incidents. Using up-to-date technology, cell phone applications also allow near miss incidents to be reported to a central number, eliminating the need for paperwork and eliminating any delay. Pre-programmed computer reporting systems have also been in operation for a while.

While all reports should be listed on the main database and be acknowledged, not all deserve investigation. Many reported situations would have already been rectified, but near miss incidents that have high potential for severe loss should be investigated with the same vigor as accidents are.

OCCUPATIONAL INJURY, DISEASE, AND DAMAGE STATISTICS (S4, E4.5)

Injury, illness, and property damage statistics should be compiled by the safety department for tabling at the executive and other safety committees. These statistics should preferably be in the form of pie charts and graphs, and should be calculated on a monthly basis, as well as a 12-month moving average. Statistics can be broken down to reflect statistics per department, or group within the organization, as well as the whole organization.

Many organizations keep graphical data on body parts injured during the month, which help indicate possible PPE program weaknesses. Statistics concerning hazards rectified could be an important safety performance indicator (SPI).

LOSS STATISTICS (S4, E4.6)

A person made responsible should compile a table listing all the downgrading events' costs for each month and progressive year to date. This should include all costs related to unwanted events that occurred on the premises, for example, fires, vehicle accidents, damage accidents, production losses as a result of injury accidents, medical claims, liability claims, fines for non-conformances, etc. These figures should include details of workers' compensation and insurance claims made in the preceding period. These figures should be reported to the executive safety committee at the monthly meeting.

COST OF RISK (S4, E4.7)

The total cost of risk includes costs of losses such as death and injuries caused by accidents, equipment damage, vehicle damage, and the cost of insuring residual risks, and penalty costs. The status of safety at the organization can be determined through an analysis of the cost of accidental losses. An analysis of insurance claims and accident costing will allow for the measurement of the effectiveness (or ineffectiveness) of the safety management system.

Possible rebates or reduction in insurance premiums, and elimination of high-risk penalties can motivate the costs of risk mitigation and of maintaining a safety management system. The Risk Management department usually calculates these figures.

NEAR MISS INCIDENT AND ACCIDENT RECALL (S4, E4.8)

Near miss incidents (events with no visible outcome or resultant injury or damage) are often not regarded as important. They are not seen or not recognized by the average employee, as there is no resultant injury or damage, therefore they are not reported nor investigated.

TIP OF THE ICEBERG

Numerous safety systems are injury prevention programs and only concentrate on the serious and minor injuries. This is only the tip of the iceberg. According to the book *Practical Loss Control Leadership*, it has been estimated that for every 641 undesired occurrences, 1 will result in serious injury, 10 will result in minor injury, 30 will end up causing property and equipment damage, and some 600 will have no visible outcome or consequence (near miss incidents). (Frank E. Bird, 1992) (p. 21)

LEARNING FROM PAST EXPERIENCE

Many of these unreported events are recalled by employees once the opportunity is provided, and that is the purpose of making near miss incident and accident recall, a part of the safety system. The objective of this element is to systematically gather information and learn from near miss incidents and accidents that may not have been reported, so that the information can be shared and future events can be prevented.

NEAR MISS INCIDENT

The difference between an accident and a near miss incident is purely a matter of chance, as the outcome of an undesired event cannot be determined, and is very difficult to predict. Recalling the past to improve the future is of vital importance in a safety system, and the reason that many systems fail is that the underlying problems, in the form of near miss incidents are never reported, investigated, and eradicated.

ACCIDENT AND NEAR MISS INCIDENT RECALL

Accident recall is a method of recalling past injury and/or damage causing accidents and bringing about an awareness of their causes, in order to ensure that steps are in place, so that a recurrence does not happen.

Near miss incident recall is a method of recalling near miss incidents (reported and not reported) that did *not* result in any visible injury, damage, or production loss, but which may have if circumstances were different. This is to prevent the event from happening again in the future.

PROCEDURE FOR RECALL

Accident and near miss incident recall should be a fixed item on the agenda of the safety committees, and all attendees should be encouraged to recall unreported events that could have resulted in losses. These could be work, home, or traffic related events not necessarily related to the worksite.

Five-minute safety talks should be used to disseminate information about accidents and near misses experienced in other departments of the company. Formal accident and near miss incident recall sessions should be planned and held on a regular basis, and accident and near miss incident recall should be used during all safety training courses and workshops.

Managers, supervisors, and contractors should be accountable for fully supporting the formulation, administration, implementation, and performance of this standard. The safety department should be responsible for the administration of this standard, as well as for record keeping of documentation, and distribution of safety reports and statistics concerning near miss incident and accident recall. Employees should be encouraged to participate in the daily and monthly recall sessions.

RETURN TO WORK PROGRAM (S4, E4.9)

A Return to Work Program is the practice of bringing injured employees back to work, as soon as they are medically able, to a position within the organization compatible with any physical restrictions they may have. The prompt return of injured employees to positions within their medical restrictions will help minimize the impact of work related injuries. Returning to work early helps employees remain functional as they recover, while providing the organization with the valuable use of employees' skills. It also helps control workers' compensation costs.

The organization should have a written standard for the Return to Work Program, listing all the requirements of the program as well as responsibilities and accountabilities for managing the program.

13 Work Environment Conditions

EXAMPLE SMS: SECTION 5

Section 5 of the Example Safety Management System (SMS) covers the physical work environment, its structures, appearance, and order. A safety management system can only operate efficiently if the work area reflects good health and safety conditions, as this helps influence employees' behaviors and creates a sense of business order. The following 10 elements should be incorporated under this section as a minimum (Figure 13.1).

BUSINESS ORDER (GOOD HOUSEKEEPING) (SECTION 5, ELEMENT 5.1) (S5, E5.1)

A basic element of a SMS, business order, or good housekeeping, creates a firm foundation on which all other aspects of the SMS can be built. Good housekeeping is good management, as it eliminates hazards, reduces risks, and creates a safe and clean work environment.

Once a decision has been taken to implement a world's best practice SMS, the workplace environment needs to be brought up to standard. Many factories, assembly plants, workplaces, and mines lack business order. *Business order* is defined as *a place for everything and everything in its place, always.* The workplace housekeeping needs to be clean, neat, and orderly. Even heavy industry and underground mining work areas can be immaculate, and have to be, if risks are to be eliminated and employees are to work safely.

INDICATION OF SAFETY

Business order contributes a great deal to the SMS, as it is a tangible element. Once employees see the improvements brought about by good order, they are encouraged to further improve safety at their workplaces. The standard of housekeeping at a plant gives a clear indication of the standard of safety at that plant, and is a direct reflection of management's concern for safety, and reflects employee involvement and motivation in safety. A work area that is clean, with everything in its place, and with no slip, trip, bump, or fall hazards, is an area in which workers are less likely to be injured and fires are less likely to occur.

Substandard business order (poor housekeeping), indicates the lack of an integrated safety management system, inadequate safety and health standards, low standards, or non-conformance to safety standards.

Number	Element Title	Number	Element Title
5.1	Business Order (Good Housekeeping)	5.6	Buildings and Floors: Clean and in a Good State of Repair
5.2	Aisles, Storage, Work Areas Demarcated	5.7	Good Lighting: Natural and Artificial
5.3	Scrap and Refuse Removal System	5.8	Ventilation: Natural and Artificial
5.4	Good Stacking and Storage Practices	5.9	Plant Hygiene Facilities
5.5	Color Coding: Plant and Pipelines	5.10	Pollution: Air, Ground and Water

FIGURE 13.1   Section 5 of the Example SMS has 10 elements.

BENEFITS OF ORDER

Accidents are reduced by keeping the walkways, work, and storage areas free from superfluous material. Fire hazards are reduced by ensuring that combustible material is correctly stored, and that chemicals, and other flammable substances do not come into contact with sources of ignition and other combustibles. Productivity is improved dramatically when good business order prevails, and a constant, safe and efficient flow of workers, machinery, and materials is facilitated.

Employees should be made responsible for housekeeping in their areas, and regular housekeeping inspections should be conducted as part of the safety system. Safety Representatives' inspection checklists should include housekeeping as an item to be checked. Good housekeeping competitions also help in maintaining a high standard of business order.

DEMARCATION OF AISLES, WALKWAYS, STORAGE, AND WORK AREAS (S5, E5.2)

Demarcation, or delineation, is when guidance lines are painted or taped onto the floor to indicate safe walkways, aisles, traffic ways, work and storage areas. Demarcation enables walkways to be positioned in such a way that obstacles, i.e., doors, open windows, and protruding materials do not interfere with pedestrian flow and vehicle traffic. Material stacking and storage areas are demarcated so that items are stacked in the correct places.

AISLES AND WALKWAYS

One of the objectives of workplace demarcation is to ensure that aisles, walkways, and roadways for pedestrian and vehicle traffic are well planned, so that access ways follow a logical route and are demarcated to ensure safe and efficient traffic flow.

STACKING AND STORAGE

To ensure good business order, all stacked or stored material should occupy authorized stacking and storage areas which should be demarcated accordingly. Unauthorized storage leads to poor order and bad housekeeping practices. Demarcation should separate work, walk, and storage areas. Temporary stacking should also be demarcated.

FIRE AND EMERGENCY EQUIPMENT

Emergency equipment should have "keep clear" zones demarcated below them to ensure that goods are not stored in front of the equipment. Emergency equipment must also be easily accessible at all times. Examples are: first aid stations, emergency eyewashes, emergency showers, etc. Fire equipment needs to be accessed immediately in an emergency and should be freely available and not obstructed. Demarcation of "keep clear" areas around the equipment will indicate that items are not to be placed, or stored, in these keep free zones. All emergency equipment should be suitably demarcated and their position indicated by correct signage posted above the equipment (where practicable). Regular inspections should take place to make sure that the demarcated areas are free from obstruction.

TRAFFIC DEMARCATION

Roads, parking areas, and travel ways used by vehicular traffic should conform to the country's traffic demarcation codes. Standard road marking should be applied for roads and parking bays, and traffic signs must be according to code. Speed limits should also be posted on company roads and the same rules of the road that apply outside the company property, should apply on the property.

SCRAP, WASTE, AND REFUSE REMOVAL (S5, E5.3)

TRASH BINS

Trash bins for the disposal of trash materials should be strategically placed throughout the workplace, including work areas, operating areas, offices and canteens, and must be present in sufficient quantities to promote good housekeeping practice. A scheduled removal system should be in operation to empty the bins at appropriate intervals. Trash bins containing food wastes should be constructed of smooth, corrosion resistant, easily cleanable materials, or lined with disposable plastic liners, and be provided with a solid, tight fitting cover when used in a designated eating area.

Receptacles should be maintained in a clean and sanitary condition. Trash bins could be color coded and the word *Trash* could be stenciled on the side of the bin in 4" (10 cm) black letters, or suitable decal stickers could be used. Separate metal bins should be available for disposal of rags and waste that are contaminated with combustible materials. All contaminated rags should be deposited in these bins immediately after use. These trash bins should also be suitably color coded. The word *Combustibles* could also be stenciled on the side of the bin in black letters, or suitable decal stickers may be used. Combustible trash bins should be removed from the site at least once daily, or at the end of each shift.

RECYCLABLE AND SALVAGE MATERIAL BINS

Recyclable material bins should be available for the disposal of recyclable materials. These bins must be strategically placed throughout the workplace. Recyclable material bins could be painted in the applicable color code, and the type of recyclable material within the bin (e.g., paper) could be stenciled on the side of the bin.

Salvage material bins should be available for the disposal of salvage materials. These bins should also be strategically placed throughout the workplace in sufficient quantities.

REGULAR, CONTROLLED REMOVAL

All bins should be emptied on a regular, scheduled basis. Supervisors should keep a schedule for bin removal and be able to demonstrate that the schedule is being followed. Following waste disposal, empty bins should be returned to their approved locations.

DEMARCATION

All waste bins should be kept in clearly demarcated locations. The position of trash cans, drums, containers, or dumpsters should be demarcated with an OSHA Safety Yellow (or the company's own color code), square or rectangle, on the floor, below the bin, by means of demarcation lines. It is the responsibility of the work area supervisor to ensure that demarcation lines are not allowed to deteriorate to the extent that they become indistinct.

STACKING AND STORAGE PRACTICES (S5, E5.4)

Stacking and storage of equipment, goods, and materials should be done in such a way that an hazard-free environment, which conforms to the company's housekeeping standards, is created and maintained. Proper stacking and storage is essential when establishing and maintaining a safe work environment. Correct safe stacking and storage practices prevent damage to equipment and materials, and injury to persons. Proper stacking and storage ensure maximum utilization of space and reduces production time lost due to searching for supplies and materials.

RESPONSIBILITY AND ACCOUNTABILITY

The manager, supervisor, or contractor (Responsible Person) of the area is accountable for the stacking and storage within his or her area of responsibility, and should rectify any deviations to ensure compliance with the standard. Where applicable, an experienced person should be appointed to supervise stacking practices.

INSPECTIONS

Responsible Persons, safety personnel, and Safety Representatives should carry out monthly inspections, and any deviation noted concerning stacking and storage must be rectified. The safety department is responsible for conducting inspections on an ongoing basis, for reporting deviations, and initiating the actions required to maintain compliance with stacking and storage standards.

COLOR CODING: PLANT AND PIPELINES (S5, E5.5)

The objective and scope of this safety system standard is to establish a common color coding system that allows for speedy recognition, which warns employees of hazards, and therefore lessens the chance of error. There is no accepted international standard for color coding, even though some countries have national standards. Some industries have their own unique color coding system. The use of the following colors is intended as a guideline only.

YELLOW

OSHA safety yellow could be the basic color for designating caution and for marking physical hazards such as:

- Striking against, slip, trip or fall, and "caught in between."
- Objects in places where caution should be exercised.
- Radiation hazard areas.
- Housekeeping, stacking, and storage demarcation.
- Yellow and black diagonal stripes can be used to demarcate bump against hazards.
- No parking areas and walkways.
- Locations of explosive substances.
- Disposal cans for contaminated metal.
- Rooms and areas where radioactive materials are stored.

RED

Signal red can also be used with white lettering or stripes depending on the application. Among other applications, red can be used to indicate the following:

- Safety cans or other portable containers of flammable liquids having a flash point at or below 80°F (27°C) should be painted red.
- Stop buttons, or electrical switches on which letters or other markings appear, used for emergency stopping of machinery should be red.
- Firefighting and protection equipment such as fire extinguishers, fire lines, and other fire appliances.
- Fire protection materials and equipment. This classification includes sprinkler systems and other firefighting or fire protection equipment.
- Red may also be used to identify or locate such equipment as alarms, extinguishers, fire blankets, fire doors, hose connections, hydrants, and any other firefighting equipment.
- Red can be used to indicate "Danger" when blasting is in progress.

GREEN

Emerald green, in conjunction with stripes or edging can be used to indicate the following:

- The location of safety and first aid equipment
- Emergency exists and safety areas
- Information signs
- Starting devices on electrical equipment
- Miscellaneous safe conditions

ORANGE

Orange can be used on the following:

- Electrical switch gear
- Electrical services
- Exposed and rotating machine parts
- Inside of machine guards
- Conduits and live (hot) fittings

PIPELINE COLOR CODE

Pipeline color codes, or labeling systems, are used to indicate the content of a pipeline, the direction of flow, and other important information. Positive identification of a piping system's contents could be via means of a basic color with color bands to indicate different contents. Legend boards to interpret the color banding system should be placed at strategic places within the plant to aid with the correct identification of the contents of the pipelines.

The contents of the pipelines could also be indicated by means of written labels attached to the pipe which give the name of the content in full or abbreviated form. Arrows should be used to indicate the direction of flow. The color coding and labeling systems should be constant throughout the plant.

STRUCTURES, BUILDINGS, FLOORS, AND OPENINGS (S5, E5.6)

The objective of this standard is to ensure that buildings and floor areas, inside and outside, utilized for business processes, are appropriate for the needs and pose no hazards to the work performed. Areas of responsibility should be designated to appropriate persons to ensure the detection and rectification of deviations. Fixed structures are to be included in the planned maintenance program to ensure their ongoing integrity. Each manager and supervisor should be responsible for the application of this standard in their work area.

BUILDINGS

Buildings should be appropriate for the work performed in them and shall allow for adequate and safe movement of people, equipment, materials, and appropriate vehicles. They should meet local construction specifications and any other legal requirements applicable, and should be designed to withstand the likely weather conditions of the region. Roofing, walls and support columns shall be free from damage and cracks that could affect their structural integrity. Gutters and downpipes should be adequate for weather needs and be maintained free of blockages.

Doors and doorframes should not be damaged and should maintain adequate support of the structure as well as sealing from the environment. Any wall, door, or structure that forms part of any passive fire protection system shall not be breached or compromised. All services passing through the barrier shall be suitably fire stopped and sealed. Windows and frames should be of appropriate strength, and broken or cracked panes replaced. Any glazing used in doorways or near traffic ways, should be safety glass. Fences and gates should be sound and should be maintained and kept free of hazards such as loose wires, etc.

FLOORS

All floors should be of sufficient structural strength to maintain the maximum working load that they may be subjected to. Floor surfaces and carpeting shall be free of holes, and uneven, or unsafe surfaces that pose a risk to people and mobile equipment. Roadways and walkways should be even, unobstructed, and free of debris such as nails, sharp objects, cords, pipes, etc. Where a temporary situation requires a cable or pipe to be run over a walkway or roadway, a suitable cable or pipe bridge, or ramp, should be provided and maintained for the duration of work. Drains should be maintained free from debris to ensure adequate runoff during rain and water discharges. Bathroom and change room floors should be of non-slip material.

DELEGATION OF RESPONSIBILITY

All areas, offices, warehouses, and workplaces should be allocated to nominated persons for inspection purposes. This responsibility can be designated in writing or on a plan. The name of the person responsible for inspecting the area could be displayed in a prominent place in that area. The responsible person should inspect his or her area of responsibility at least once per month and report any deviations through the checklist inspection sheet.

GOOD LIGHTING: NATURAL AND ARTIFICIAL (S5, E5.7)

ILLUMINATION

The purpose of this element of the SMS is to ensure that sufficient illumination is maintained, which will enable safe task performance in a risk-free work environment. An example of a safety system standard for workplace illumination follows:

- Lighting intensity surveys are to be conducted by a competent person, initially and after a modification/new installation is completed, which might influence the illumination. This survey covers all areas including offices and outside areas. The Industrial Hygienist (IH) coordinates the survey and corrective actions are taken on reported deviations.

- The level of illumination provided is determined by the industrial hygiene department or IH, and is in accordance with the U.S. Government Services Administration or other acceptable standard.
- Poorly illuminated areas, when identified, must be repaired according to an action plan based on risk. A follow up survey is to be done to ensure that corrective actions satisfy the requirements of this standard.
- Night and day illumination surveys must be done to cover the whole work area and workplaces where work might be carried out during hours of darkness.
- Monthly inspections on illumination and windows are to be done, covering the whole work area. Deviations/defects must be noted, reported, and repaired.
- All lights and light switches are to be numbered for identification to assist during maintenance or emergency work.
- Emergency lighting is installed as per requirements and tested by maintenance on a 6-monthly basis, and maintained where deemed necessary.
- Reflective clothing is to be used when working at night, and also during daytime in high-risk situations (such as working among traffic) to obtain maximum visibility where necessary.
- Road barricades and bollards must have reflective warnings or must be illuminated during night time.

VENTILATION: NATURAL AND ARTIFICIAL (S5, E5.8)

VENTILATION AND AIR QUALITY

In some industries, such as underground mining, the quality and quantity of the ventilation supplied to the underground workings is one of the most critical processes of the SMS. Adequate ventilation, heating, and cooling pose a high risk to employees if not installed and maintained correctly. This element is normally managed by specialists in the field, but the safety system should determine the standards for ongoing measurements and evaluation of ventilation. Correct instruments should be used for air quality testing, and these should be calibrated according to the manufacturer's specifications.

This element also relates to the elements of confined space entry, office ergonomics, welding and cutting, hazardous substance control, and laboratory safety, among others.

PLANT HYGIENE FACILITIES (S5, E5.9)

HYGIENE AMENITIES

Facilities provided for employees such as restrooms, bathrooms, change rooms, and showers, lunch rooms, emergency showers and eyewashes, must be adequate and maintained in a clean and hygienic state. The manager, supervisor, or contractor (Responsible Person), should be made responsible and accountable for providing clean and hygienic facilities, in accordance with the organization's standards. They are responsible for ensuring that the facilities are cleaned regularly. The supervisor, or Safety Representative is responsible for the regular monthly inspection of facilities and the results must be recorded on the relevant checklists.

Bathrooms, Change Rooms, and Restrooms

The floors of bathrooms, restrooms, and change rooms should be swept and wet mopped using disinfectant cleaners each shift, where the facility is used on a 24-hour basis. Daily cleaning would be required at other locations. Walls should be cleaned weekly, or more frequently if necessary. Windows should be cleaned monthly and mirrors are to be cleaned weekly or more frequently, if needed. Lights and diffusers should be cleaned quarterly or as required, and air conditioners should be cleaned and serviced regularly.

Toilets and Urinals

In rooms with multiple toilets, the toilets must be separated by partitions with doors. In rooms with multiple urinals, the urinals should be separated from one another by partitions, and the door to a toilet cubicle is to be provided with an interior lock. Signs indicating male/female toilets must be posted on the door, where applicable. Toilet bowls, seats, and external surfaces should be cleaned using disinfectant cleaners each operating shift where the facility is used on a 24-hour basis. Daily cleaning is required at other locations. Toilet paper must be provided at each toilet.

Hand and Face Washing Facilities

Hot and cold water should be supplied to each sink or basin and the interior and exterior basin surfaces, and faucets should be cleaned using disinfectant cleaners on each operating shift, where the facility is used on a 24-hour basis. Taps are to be in good repair and should not leak. Soap should be provided at each sink and paper towels or warm air blowers should be readily accessible.

Trash Containers

Open or covered containers should be provided for paper towel disposal. Toilet cubicles used by females must have at least one covered container, where applicable, and containers must be free of leaks and not absorb liquids, and ideally be equipped with plastic liners.

Lunch Rooms and Cafeteria

Floors of these areas should be swept and wet mopped using disinfectant cleaners daily in facilities that are used on a 24-hour basis. Walls should be cleaned weekly, or more frequently. Windows should be cleaned monthly where possible, and lights, diffusers, and reflectors cleaned quarterly. Air conditioners are required to be cleaned and serviced annually, and filters are to be cleaned as required. Ventilation exhaust fans should be cleaned and serviced annually, or more frequently if necessary, and filters of extraction hoods cleaned frequently. An adequate number of containers are to be made available for trash disposal at lunch rooms. Fridges, ovens, microwave ovens and stoves should be clean and hygienic and inspected at least once a week.

Kitchens

Kitchen cupboards and working surfaces should be kept clean and tidy using disinfectant cleaners. All washing facilities should be washed frequently using disinfectant cleaners and cleaning materials should be stored in cupboards separate from food. The control over these areas would fall under the food safety program as well.

Water Dispensers

Drinking fountains located in dusty areas must be cleaned regularly, and must be connected to a potable water supply. The fountain's water stream should be free from contact with other materials, and they should be cleaned on each operating shift where they are used on a 24-hour basis, and daily at other locations. Dirt, rust, and micro biocide filters are to be changed out at the frequency recommended by the manufacturer.

Emergency Showers

Emergency showers should be installed as close to the hazard as possible, as required by the risk assessment. Each emergency shower should have an identification number and a sign indicating "emergency shower." The shower should be correctly color coded. Water supply valves should stay in the open ("on") position and showers should be cleaned and tested monthly. The water temperature should be maintained between 68°F–86°F (20°C–30°C).

Emergency Eyewashes

A sign reading "emergency eyewash" should be erected at each eyewash station, and correct color coding should be provided. The signs should be visible and legible, and protective covers should be installed over the eyewash spigots (nozzles). Eyewashes should be cleaned and tested weekly and records kept. Water at the nozzles should be at high volume and low pressure. Eyewash station positions should be demarcated and kept free from obstruction to allow for easy access in an emergency.

POLLUTION: AIR, GROUND, AND WATER (S5, E5.10)

This element is added to the Example SMS for organizations that do not need, and do not have a comprehensive Environmental Management System (EMS), due to the nature of their business. The element is intended to cover the basic environmental requirements.

OBJECTIVE

The objective of this standard is to ensure that appropriate systems and facilities are established to monitor and control pollution of air, ground, and water in the working environment, and to meet the regulatory emission and discharge limits for various pollutants, including solid and liquid wastes.

RESPONSIBILITIES

Managers, supervisors, and contractors (Responsible Persons), shall be responsible to ensure that the release of all types of pollutants to air, water, and ground, from their activities, operations, products, and services is controlled, and all emissions, discharges, and releases are kept below the regulatory limits through implementation of management systems designed to measure and drive continual improvement in environmental performance.

Management should establish measurable objectives, targets, and key performance indicators (KPIs) throughout the plant, to ensure that the emissions, discharges, and disposals comply with the regulatory limits and standards as a minimum.

STANDARDS

If the scoping study on a new project recommends it, a comprehensive environmental impact assessment study (EIA) should be carried out to identify the total pollution potential of the project and appropriate environmental management plans (EMPs) should be prepared and implemented to reduce the pollution load to a practicable minimum.

A system should be in place to identify and provide access to all applicable environmental laws, regulations, approvals, licenses, permits, and other requirements such as codes, charters, standards, commitments, etc. These should be documented in a register of legal and other requirements.

Where local legislation does not require an adequate level of environmental performance, activities should be conducted in a manner that is consistent with relevant international standards and practices, taking due account of social and cultural sensitivities.

Systems should be in place to ensure that environmental documents are established, (accurate, current, legible, and identifiable with revision dates), and are periodically reviewed, updated, and approved. Obsolete documents should be promptly removed, or otherwise assured against unintended use.

14 Safety Management System Implementation Strategy

SAFETY CONTROL

Safety controlling is what takes place when a Safety Management System (SMS) is implemented. This control comprises the establishment of safety and health standards, implementing them, and monitoring adherence to them. These standards form the safety management system which should be integrated into the daily management of the organization. The actions and processes called for by the standards are ongoing and will eventually lead to a minimization of risks within the organization.

DEFINITION

Safety control is the management function of identifying what must be done for safety, setting standards of measurement and accountability, inspecting to verify completion of work, evaluating, and following up with safety action.

The identification phase would include highlighting the strengths and weaknesses in the existing safety system if any, *identifying* the risks in the workplace and *identifying* what actions need to be taken, by whom, and when, to set up the safety system. The next step is *setting standards of performance* and *standards of accountability* for safety system processes. The *measurement* against those standards is then done by *inspections* and follow up. The *evaluation* of work being done is verified and evaluated by audit, and *corrective actions* put into action to correct deficiencies.

A modern safety approach suggested by some Guidelines is the *Plan, Do, Check, Act* methodology.

- *Planning* is the establishment of objectives, processes, and actions necessary to deliver results in accordance with the safety policy leading statement. (Philosophy, policy statement, action plan, setting objectives, etc.)
- *Doing* entails implementing the processes and actions to achieve the goals and ambitions set by the guiding policy document, and relevant standards. (Setting standards, training, involvement, baseline audits, implementing standards, etc.)
- *Checking* is the monitoring phase where achievements are gauged against the policy and standards. (Inspections, reviews, system audits, measurements, etc.)

- *Acting* is taking continual action to improve the safety system and its processes. (Implement more elements as per action plan, rectify deviations, recognize performance, etc.)

When implementing (and maintaining) a structured safety management system, the *Plan, Do, Check, Act* methodology is a good model to follow, as it leads to a process of continual improvement, which is important in safety management. A safety system has no end point, and management will continually be challenged with ongoing processes and improvements.

DEFINING THE ORGANIZATION'S SAFETY PHILOSOPHY

One of the first (planning) stages in the implementation of a safety system is a firm commitment and decision from top management to implement and support the safety initiative, and to remain active participants in the change process (Figure 14.1). Once this commitment has been made, the organization must establish a safety and health philosophy. This does not have to be a long complicated document, but rather a few simple safety beliefs that the organization believes in, and will abide by.

An example of an organization's safety and health philosophy is:

- Our employees are our greatest assets.
- We believe that more than 98% of accidents can be prevented.
- Each accident is caused by multiple causes.
- Contractors are as important as our employees.
- Safety is a part of our daily business.
- We believe in safe production, etc.

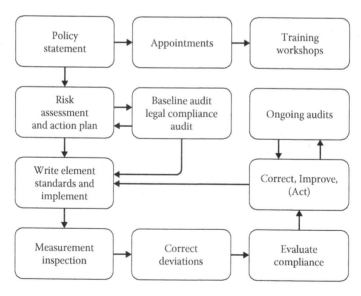

FIGURE 14.1   Safety system implementation sequence.

A few guiding statements such as those listed will establish a firm premise on which to build the safety system. Organizations should avoid old, worn out clichés such as "Safety is first" or "Safety is our priority," in their philosophy as they are meaningless, and do not reflect reality. The safety and health policy is now constituted around the safety philosophy.

THE SAFETY AND HEALTH POLICY STATEMENT

Most safety and health management systems start at a logical point, which is a safety and health policy statement of intent (safety policy) which is initiated, and followed through, by management. This is the guiding document of the safety management system and forms an important step in the planning phase.

A safety policy is a standing safety decision, which applies to repetitive safety problems that may affect the safety of the organization. It is the commitment that the management team makes to guide the safety effort and activities.

The following are advantages of a safety policy:

- The policy gives common points of view concerning occupational safety and health.
- The policy provides rational rather than erratic safety decisions.
- Employees know the executives' stance on safety.
- The policy is an indication of the organization's safety considerations.
- Safety policies allow for delegation of safety work.
- Efficient and effective teamwork is facilitated by a safety policy.
- The policy provides guidelines to everybody in how to do the right things concerning safety.
- The safety policy is managements' commitment to safety and the safety system.

Safety policy statements must be dynamic, they must be realistic, and the objectives set must be tangible. Depending on the organization, both management and worker representation should agree upon the policy. The Safety and Health policy should be extensively publicized, should cover all aspects of safety, and should:

- Include a commitment to injury and ill-health prevention
- Indicate safety responsibilities and accountabilities
- Refer to continuing improvement of safety and health initiatives
- Contain a commitment to comply with safety and health legislation
- Provide a framework for safety objective setting
- Be documented, displayed and maintained
- Be freely available
- Be communicated to all affected parties
- Receive periodic review

PUBLICIZING THE POLICY

The safety policy should be drawn up with considerable care, and must be signed by the Chief Operating Officer or the most senior manager. In some instances, the entire executive leadership signs the policy. The safety policy should be displayed at prominent positions throughout the organization. To give credibility to safety, the policy should preferably be attractively printed and suitably framed. The display positions must be carefully selected, and employees should be briefed and reminded of the policy via means of ongoing briefing sessions, to ensure that everyone understands the policy. It is advantageous to have a copy of the safety policy reproduced in the safety rule book. This helps during the safety induction process and also serves as an ongoing reminder of the safety commitment of the organization. Contractors should also be made aware of this policy from the beginning of the bid process.

ACTION PLAN

An important step in implementing a safety management system is the compiling and issuing of a safety management system implementation plan. This plan will then form the roadmap for the laying out of the safety management system. The plan will detail what needs to be done, by when it must be completed, and who-must-do-what, to complete the plan objective. The plan will consist of a number of small objectives to be achieved and will lead to the complete implementation of a well-structured safety management system, over a predetermined period of time.

TIME SPAN

Few, if any, safety management systems are implemented in a period shorter than 5 years. This is dependent on the level of safety management existing at the initiation of the structured safety system launch. Complying with local safety laws does not indicate the implementation of a world's best practice safety management system. Legal compliance is seen as a minimum requirement. If there is uncertainty as to the status of the safety management system, the baseline audit will clarify further enhancements and existing weaknesses.

The management review recommended by the Guidelines would consist of frequently comparing the safety management system progress with the action plan.

GUIDELINE CHOICE

As with all strategies, there need to be objectives, and achieving the requirements of the selected safety system Guideline could be one of them. This would ensure that the safety management system is following a national or international standard for safety management systems. An advantage of aiming to achieve the requirements of a Guideline is that external third party audits can be facilitated to indicate to the organization whether or not they are fulfilling the Guideline's requirements. In some instances, it leads to accreditation.

MANAGEMENT AND EMPLOYEE AWARENESS TRAINING

Few managers have undergone formal safety and health training, and therefore do not fully understand the functioning of a safety system, nor the philosophy behind loss causation and safety control. A series of workshops should take place where the organization's safety philosophy, policy, and safety action plan are presented to all line management. Their roles, authority, responsibility, and accountabilities should be part of the presentation. Many organizations insist that ALL their managers, supervisors, and team leaders attend these workshops. The same training workshops should be held for employees so that all are made aware of the policy and the safety system requirements. Contractors should also be exposed to similar workshops.

INTEGRATING SAFETY AND HEALTH

Safety and health should be integrated into the daily routine of the business. All meetings should open with a discussion on safety and incident recall. Ongoing safety processes should form part of the normal work, and the message must be clear that safety is not simply an add-on. The policy must be lived. Line managers must set an example to all, and employees need to be engaged in the safety processes and activities.

HAZARD IDENTIFICATION, ELIMINATION, AND RISK ASSESSMENT

The risks within the organization will indicate the controls that need to be put in place via the safety management system. All hazards need to be first identified and the hierarchy of controls applied in an effort to reduce the hazards and the consequent risks.

THE HAZARD BURDEN

The range of activities that take place at an organization and the nature of the process will create hazards, which will vary in nature. The number, range, nature, distribution, and significance of the hazards, known as the hazard burden, will determine the risks which need to be controlled. Ideally, the hazard should be eliminated altogether, either by the introduction of inherently safer processes, or by no longer carrying out a particular activity. If the hazard burden is reduced and if other variables remain constant, including the successful functioning of the safety management system, this will result in lower overall risk and a consequent reduction of unintentional losses.

HAZARD ELIMINATION

The first option to reduce hazards is to endeavor to eliminate them completely. If the hazard could be designed out of the workplace, this would be the ideal solution.

If elimination is not possible, then engineering controls such as guarding or barricading to isolate the source of energy would be the next option. Guarding of rotating machinery and other pinch points is a good example of isolating the source of the hazard. If hazards cannot be eliminated by engineering revision, then the next option is substitution of the process, substance, or procedure with another safer process, substance, or procedure.

Administrative controls are applied when the above options have been exhausted. These would involve actions, such as minimization of employee exposure to the hazard, training, awareness campaigns, work procedures and practices, permit system, etc.

Providing personal protective equipment (PPE) is seen as the last resort to protect workers. It is always more beneficial to ensure that the work environment is free from hazards before simply issuing protective clothing and equipment. PPE slows down the process and relies on the integrity of the employee to wear it, and does not eliminate the hazard. It only offers protection against the hazard, and therefore is deemed the last solution for protection against hazards.

RISK ASSESSMENT INSPECTIONS

To identify and mitigate the risks, management, in conjunction with the workforce, should visit the point of action, carry out inspections, hold discussions, and determine exactly what work needs to be done to make the work environment safer and healthier. Listening to the workers is invaluable in this regard. A baseline safety audit would assist this process. The identification of work to be done to train, guide, educate, and motivate workers toward safe work practices should also be considered.

SAFETY MANAGEMENT SYSTEM

Based on the risk assessment, management lists and schedules the work needed to be done to create a safe and healthy work environment, and to eliminate high-risk acts of employees. This would mean the introduction of a suitable structured safety management system based on world's best practice. All safety management systems should be based on the nature of the business and be risk-based, management-led, and audit-driven. The introduction of a structured safety system could include having to guard machinery, demarcate walkways and work areas, purchase correct tools and equipment, and set up maintenance systems for equipment and plants, compile procedures, initiate processes, etc.

CHANGE AGENT

Often large corporations hire a safety culture change agent as a part of their implementation strategy. By creating a safe space, the agent holds discussions with employees at all levels, and through a series of discussion and feedback sessions, he or she determines what safety changes the employees themselves can make. This process is called the empowering of employees in safety. The same process is then applied with members of line management, and at the end, both groups have a clearer picture of the role they can play in the safety system.

BASELINE AUDIT

A baseline audit could be used as a starting point for the implementation of a structured safety system. The baseline audit would deliver a report indicating strengths and weaknesses compared to a world's best safety system. An action plan compiled as a result of the baseline audit would then form the safety system implementation plan.

LEGAL COMPLIANCE

Using the local safety regulations as a basis for an organization's standards is always a recommendation. Although regarded as the minimum to strive for, they offer already written standards that only have to be converted to the organization's specific needs and customized. One of the objectives in the safety policy is legal compliance, so this would be a great help in establishing system standards.

SAFETY MANAGEMENT SYSTEM STANDARDS

The next logical step in the implementation process is to set written standards of performance. Safety standards are referred to as "measurable management performances." Standards are set for the level of work to be done to maintain a safe and healthy environment, free from actual and potential accidental loss. Standards are established in writing for all the safety and health management system elements. Without standards, the safety management system has no direction, nor are safety expectations established. (If you do not know where you are going, any road will take you there.) Setting standards for the safety of the workplace will entail determining the major areas of risk, and in conjunction with various committees, unions, and employees, drawing up acceptable standards for at least eighty elements, similar to the Example SMS.

Many of these standards have already been defined and written by prominent safety organizations in various countries, so the safety wheel need not be re-invented. Local safety and health regulations also prescribe certain standards and these could be used as a guideline. Many years of research and input from numerous quarters have contributed to the development of these standards.

Management should modify these standards to suit their own company's requirements but should not water down, or eliminate the standards merely, because they are too difficult to achieve. These should be regarded as the minimum standard to achieve.

The standards should be measurable management criteria, and therefore they should be tangible, reasonable, attainable, and quantifiable. Standards would also include time frames for initial completion and ongoing updates. Safety system standards must be based on safety control and not consequence. Any standard set should include *what* must be done, *who* must do it, and by *when* and *how often* must it be done.

SET STANDARDS OF RESPONSIBILITY AND ACCOUNTABILITY

Standards of accountability are now set by delegating authority to certain positions for ongoing safety work to be done. Coordination and management of the safety system needs to be allocated to certain departments and individuals, and this standard dictates who must do what, and by when, to run and maintain the system.

Traditionally, the function of safety was dumped in the safety department and they were told to manage the safety. Safety does not belong in the safety department, but with line management.

Changing Safety's Paradigms explains:

> Perhaps one of safety's biggest stumbling blocks is the responsibility for safety has been pushed down to the safety department. As soon as there is a safety issue, it becomes the responsibility of the safety department. Safety belongs with the line management, from the lowest level of management to the chief executive officer. Safety is their function. The safety department should only coordinate the safety activities and not accept responsibility for the entire safety function. (McKinnon, Ron, 2007, p 135)

Setting standards of accountability is where the management team states who must do the work and by when. This would entail appointing certain responsible persons such as the following:

- Managers
- Safety Coordinator
- Occupational Hygiene Coordinator
- Safety and Health Representatives
- Divisional safety committees
- Central safety committees
- Accident investigators
- Responsible engineers
- Internal safety auditors
- Fire Coordinator
- Permit issuers
- First aid / responder attendants
- Housekeeping coordinators, etc.

Company letters of appointment, as well as a brief description of their responsibilities and accountabilities, should be issued to the nominated employees. The correct person must be appointed for the task, and adequate, relevant training must go hand in hand with this person's appointment. The responsibilities should also be included in each safety system element standard.

The nomination of safety committee members should be done on a democratic basis and the chairperson likewise elected, where applicable. The most important committees to be established under these standards would be the joint management and union safety and health committees. These committees would assist in the setting of standards of accountability of the various persons. It also offers a platform for open and frank safety discussion between management and employee representatives.

When setting standards of accountability, it should be remembered that accountability for safety can never be delegated. The individual worker remains ultimately responsible for their own safety. Managers are responsible for the employees working under them, and within their area of control. Employees such as the Safety Coordinator are only responsible to coordinate the safety system and can never be held responsible for the safety of the organization. Again it is stated, that safety is not the responsibility of the safety department. The safety department has seldom the authority to carry the burden of safety responsibility. They can only be responsible for coordinating the activities that constitute the ongoing safety system.

The management work of setting standards of accountability is of utmost importance to the success of the organization. If permit systems are needed, for example, then employees must be given the responsibility to devise these systems, modify them, and implement them. This gives them the satisfaction of helping to achieve a high standard of safety.

Some assistance can be achieved by:

- Visiting other plants that run a successful safety system.
- Calling in safety consultants to help write safety standards.
- Contacting local safety organizations to assist.
- Liaising with the local legal inspectors for advice concerning standards.

IMPLEMENTATION OF STANDARDS

Some standards may already be in place, and others may have to be implemented. The decision on which standards to implement first, will depend on the situation and other factors, but experience has shown that the smaller the safety change, the smaller the resistance to that change will be. Gradual implementation of the tangible elements of the system is preferable. Resistance to these changes should be expected and pushback from both managers and employees should be handled correctly by endeavoring to get them to be part of the change. This is why participation and recognition are important safety management functions. Gradual implementation of new safety processes will not invoke as much resistance as instant change. Implementation of the standards for the elements should be as per the action plan, which could be a one to five-year process.

The standards' implementation process will occur over a reasonable period of time, as all 80 plus elements cannot be written and implemented simultaneously. They must be phased in over time as determined by the action plan. This is why it is beneficial to implement the easier standards first. The standards that already exist in the organization, or those required by safety regulation, should be formalized and implemented first.

INSPECTIONS

Once these standards are in place, regular inspections should take place to monitor progress against these standards. After an inspection, a checklist of work to be done to fully comply with the standards should be produced, and this would form the action plan for the next few weeks. This ongoing process should continue until the first internal audit takes place, and then continue as part of the continual improvement strategy.

CORRECTION

Deviations from safety system standards need to be prioritized and corrected. Using the simple A, B, and C hazard ranking system, repairs and rectifications of hazards noted can be scheduled. To retain credibility, management at all levels should ensure that hazards reported by employees are acknowledged and rectified. This will give employees more confidence to report hazards in the future.

FOLLOW UP

A safety system calls for ongoing improvement as a result of review processes. Follow up action must take place to ensure any gaps and weaknesses found in the system are rectified. Correct reporting lines must be established, and all reports and subsequent action need to be monitored for completion. Safety system standards may need revision before the annual review, and new standards may be needed to support new production techniques or changes in the operations.

SAFETY SYSTEM AUDITS

To monitor progress and provide improvement, a system of internal audits of the safety system should be introduced within the first 6 months of the initiation of the system. The findings of each audit should be carefully monitored, and follow up action taken to improve the deficiencies found and to build on the safety system strengths. External audits should take place every 12 months and action plans should be drawn up to ensure a cycle of continuous measurement, evaluation, and improvement of the safety system and its components.

SAFETY POLITICS

Despite all efforts, interventions, initiatives, and safety processes being implemented to set up the safety management system, internal safety politics will prove to be the biggest obstacle to overcome. Safety is 90% politics and 10% work. Safety management tests and tries all levels of leadership, and challenges them with obstacles that they have never before encountered. Safety brings about change and the majority of people are not comfortable with change. There will be reaction and resistance to safety change. The resistance may come from some management levels and even from the workforce. Safety challenges the leadership ability of managers, and since the safety system impacts on most areas within the company, it tends to attract reaction, some good, and some bad.

BLAME THE SYSTEM

Managers will tend to revert to the management styles that they know, and in times of crisis forget to apply new management principles and practices demanded by a safety management system. One department's safety improvement and success will create competition, which will earn the system discredit. Many will say the system is failing because they are struggling to meet the standards required. Some will mock the safety efforts by, for example, treating the evacuation drills as a fun exercise instead of taking these seriously. The successful implementation of a safety management system requires hard work, tenacity, and ongoing drive, and support from the leadership of the organization.

15 Measuring Performance

MEASUREMENTS OF SAFETY PERFORMANCE

Safety effort and experience can be measured in a number of ways which are; upstream, (leading indicators) or downstream (lagging indicators). Downstream measures are often termed safety performance, but are not measures of performance, but rather measures of performance failure. Leading indicators are better measures of safety performance, as they are tangible, controllable, and quantifiable items. Measurements of safety efforts are important in a structured safety system (Figure 15.1).

INJURY STATISTICS

For decades, organizations have used injury statistics to measure safety. The reason these statistics and figures are used is because the term *safety* has become synonymous with the term *injury*. These statistics, such as the lost time injury frequency rate (LTIFR), disabling injury incidence rate (DIIR), and injury severity rate, etc., are really measures of injuries and not measures of safety per se.

Safety is the control measure implemented to reduce the probability of accidental losses occurring. Injuries are only one result of poor safety. Organizations incorrectly continue to count the number of casualties and equate this to safety performance. This is not an accurate gauge of safety.

Injury rates are poor measures of safety. If under reporting in an organization is 50%, then only half of the injury rate is depicted. The safety record could improve with increasing dishonesty brought about through traditional safety competitions, incentive schemes, and disciplinary programs. These all artificially help to lower (hide) the rate. The criteria for measuring these injuries differ greatly and therefore, the end results cannot be accurately compared. The players are playing the same game to different rules.

TOO MUCH EMPHASIS

Statistics can be manipulated, and due to certain pressures are often adjusted to reflect better than actual performance. Since many antiquated methods of gauging safety performance are based on the injury count, there is pressure to reduce this figure to acceptable levels. This is called manipulation, and research has shown that nearly 90% of organizations do not present accurate injury rate figures. This is simply because too much emphasis is placed on the injury rate as a measure of safety.

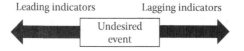

FIGURE 15.1   Leading and lagging indicators.

DOWNSTREAM MEASURES (LAGGING INDICATORS, POST-CONTACT)

There are numerous ways to record, measure, and present injury statistics, and the following are a few of the most widely used. The time period for the workhour exposure should be as long as possible to make for meaningful calculations. The injuries used for the calculation must be experienced over the same time period as the exposure. Progressive moving average calculations covering 12 months are normally used (Figure 15.2).

Disabling Injury Frequency Rate (DIFR)

A disabling injury is a work injury, including occupational diseases and illnesses, which arises out of and in the course of employment and which renders the injured person unable to carry out his/her regular established job (the job he/she normally does) on one or more full days or shifts, other than the day or shift on which he/she was injured.

The disabling injury frequency rate, or lost time injury frequency rate, records the number of disabling or lost time injuries experienced per one million employee hours of exposure. One million employee hours are used as this equates to approximately 500 workers working for one year. The formula for the disabling injury frequency rate (DIFR) is:

$$DIFR = \frac{\text{Disabling injuries} \times 1,000,000}{\text{Employee-hours worked}}$$

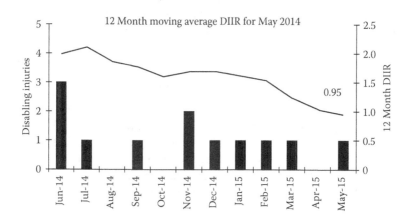

FIGURE 15.2   A 12-month moving average disabling injury incidence rate.

DISABLING INJURY INCIDENCE RATE (DIIR)

The disabling injury incidence rate (DIIR) is the number of disabling injuries experienced per 200,000 work-hours of exposure. The DIIR equates to 100 employees working for approximately 1 year, giving an exposure of 200,000 work-hours. This also represents the percentage of workers injured, and is calculated as follows:

$$\text{DIIR } (\%) = \frac{\text{Disabling injuries} \times 200,000}{\text{Employee-hours worked}}$$

DISABLING, OR LOST TIME INJURY SEVERITY RATE (DISR)

The disabling injury severity rate (DISR) is a term for measuring the actual number of days, or shifts lost per million work-hours, or per 500 employees working for a year.

Although the severity of an injury is fortuitous and therefore not an accurate or meaningful measurement of safety performance, it is nevertheless essential for costing actual losses. The disabling injury severity rate is calculated as follows:

$$\text{DISR} = \frac{\text{Shifts lost} \times 1,000,000}{\text{Employee-hours worked}}$$

DISABLING INJURY INDEX

The disabling injury index (DII) is a combination of frequency and severity into a single measure and is calculated as follows:

$$\sqrt{\text{DII}} = \frac{\text{DIFR} \times \text{DISR}}{1000}$$

(The square root of the final calculation then represents the disabling injury index.)

If this DII is to be used to determine the percentage of improvement between two DII indices, the square root of each DII must be performed before making the comparison (e.g., $\sqrt{\text{DII}} = \text{XX}\%$).

FATALITY RATE

The fatality rate is the number of fatal injuries as a result of accidents per 1000 workers, per annum. Some countries calculate per 100,000 workers per annum. The formula for calculating the fatality rate is as follows:

$$\text{Occupational fatality rate} = \frac{\text{Number of fatalities} \times 1000}{\text{Workforce}}$$

This then indicates the number of fatalities experienced per year per 1000 workers.

MILLION WORK-HOUR PERIODS

The periods between disabling injuries can also be statistically calculated in work-hour exposures, or work-shift exposures and labeled "shifts worked without disabling injury," or "work-hours worked without disabling injury." A most common objective is to work at least 1,000,000 employee hours without a disabling injury, once again equating to 500 employees at a workplace, working without injury for 1 year.

OFF-THE-JOB INJURY RATE

Off-the-job or home safety should form part of the safety system, and employees and their families should be encouraged to report off-the-job injuries, accidents, and near miss incidents. Off-the-job injury rates are normally between 3 to 6 times higher than on-the-job injury rates, and also lead to work downtime, process interruption, and loss to the company, as a result of the injured employee's absence. The off-the-job injury rate is calculated by the following formula:

$$\text{Off-the-job injury rate} = \frac{\text{Off-the-job injuries} \times 1,000,000}{321 \times \text{Number of family members}}$$

The 321 represents the number of hours each member spends at home during a month and the final figure is also represented as off-the-job injuries per million hour's exposure to home hazards.

NUMBER OF SHIFTS LOST

The number of shifts lost is a figure, which is calculated progressively and presented on a month-to-date progressive basis, as well as a monthly basis, for comparative purposes. When doing accident costing, it should be remembered that a shift lost represents a significant loss in potential income, and not only a loss of the employee's wages. For example, if an employee is paid $500.00 per day, he is normally charged out to customers at $1500–$3500 per day, and the billing figure of $3500.00 should be used to calculate the costs per shift lost.

FATALITY FREE SHIFTS

The measurement of the number of shifts without a fatality is popular among the mining industry, and is the number of worker shifts worked without a fatality. The fatality free shifts could be for an underground section, a surface workshop area and services complex, or the entire mine or plant. The fatality free shifts measurement is merely expressed as the number of shifts "exposure" which have been worked without experiencing a fatal injury.

BODY PARTS

Parts of the body being injured as a result of accidents should also be compiled and tabulated. This could include the number of head injuries, chest injuries, arm injuries, finger injuries, etc., and will provide valuable information to management.

The body parts that are being injured more frequently will then indicate a problem area, and also help direct safety campaigns and the safety effort in general.

ACCIDENT RATIO

The accident ratio is a triangular model depicting the ratio between disabling, or serious injuries, minor injuries, property damage accidents, and near miss incidents. The philosophy of safety is that, had circumstances been slightly different, the property damage accident may have injured someone, and each one of the near miss incidents with no visible loss *could have* resulted in either a property damage or an injury causing accident (Figure 15.3).

The ratio should be plotted comparing the relationship between:

1. Disabling or serious injuries.
2. Minor injuries.
3. Property damage accidents.
4. Incidents with no visible loss or damage.

Once the total number of figures have been obtained, a ratio should then be calculated by dividing all the statistics by the number of disabling injuries which will give one disabling injury at the apex of the triangle. The ratio is then read as, "For every one disabling injury experienced, (x) minor injuries are occurring, (y) property damage accidents are occurring, and (z) number of near miss incidents take place."

For example:

A company experienced 14 serious injuries, 44 minor injuries were reported, 60 property damage accidents occurred, and 200 near miss incidents were reported during a specific period. This gives a total of 264 events. Divide each category by the number of serious injuries (14). This gives us a ratio per one serious injury. Dividing the serious injuries by itself (14) would give us the (1) at the top of the ratio. Dividing the number of minor injuries by 14 would equate to 3.1 minor injuries per each serious injury (44/14). Dividing the property damage accidents by 14 will equate to 4.2 property damage accidents per each serious injury (60/14), and dividing the

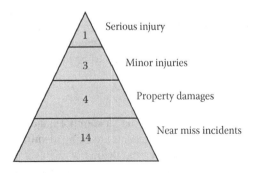

FIGURE 15.3 Calculating the accident ratio.

number of near miss incidents by 14 (200/14) will give a figure of 14.2 near miss incidents per (1) serious injury (Figure 15.3).

MOST SIGNIFICANT STATISTIC

This ratio is one of the most important statistics in any safety management system. It clearly indicates whether or not most injuries are being reported. The ratio also indicates the effectiveness of the near miss incident reporting system, and the formal and informal near miss incident recall sessions, as the number of near miss incidents should be greater than damage and injury causing accidents. The reduction of the base of this triangle would be the objective of the safety system. If the near miss incidents are reported, investigated, and remedied, the loss producing accidents should be reduced significantly.

UPSTREAM MEASURES (LEADING OR PRECONTACT INDICATORS)

Upstream measures of safety performance are actions, interventions, and processes that take place on a regular basis as part of the safety management system. They are not triggered by undesired events, but occur as part of the system. They are also referred to as precontact control measures.

SAFETY SYSTEM AUDITS

The most reliable upstream measure of safety effort is the final score delivered after a safety system audit. The score achieved is a positive indication of management work being done to combat accidental losses. All the processes audited can be controlled by management, and if functioning correctly can bring about a reduction in the probability of undesired events occurring. With the introduction of an international Guideline and standard for safety management systems, in the form of ISO 45001, organizations will be able to measure their performance against internationally accepted standards.

INSPECTIONS

The number of safety inspections conducted and hazards rectified as a result of the inspections is another measure of proactive safety processes. These inspections could include those done by the Safety Representatives on a monthly basis, but all other safety inspections, tours, or surveys should be included as well.

NEAR MISS INCIDENTS RECTIFIED

Near miss incidents reported is perhaps the most important of all the leading indicators, as it indicates that the potential accidents are being identified and rectified before the accident happens. It also indicates the degree of involvement of the workforce by the number of near misses reported. The follow up and remedial measures completed are what should be considered as a measurement figure. Near miss incidents reported without follow through and rectification would prove to be meaningless. A most important gauge of safety effort is the number of hazards that have been identified and rectified.

EMPLOYEES TRAINED

The number of employees trained (including annual refresher) in safety, health, first aid, or fire prevention is an important measure. An organization should strive to train and retrain at least 5% of its workforce annually. Refresher safety training should be attended by all employees each year to meet with best practice. Employees trained in firefighting and first aid could be included in these figures.

TOOLBOX TALKS

The number of toolbox talks and attendees is a clear indication of safety communication taking place. These should take place each morning and before the commencement of a task of work, or beginning of a shift. Targets should be set as to the number of talks that will be held during a period as well as the attendance at the talks.

SAFETY COMMITTEE MEETINGS

There should be a hierarchy of safety committees on all levels within the organization and they should meet monthly as a minimum. These numbers are a measurement of safety activity and should form part of the safety performance indicators (SPIs).

EVACUATION DRILLS

Targets should be set for the number and format of emergency evacuation drills to be held by each division. Again, these targets form part of the SPIs and should be measured monthly, and on an ongoing basis.

ELEMENT STANDARDS UPDATED

The safety system element standards need to be reviewed annually at a minimum. This review process should form one of the safety targets, and should also be used as a measurement of safety performance. This will ensure that the element standards are reviewed and modified as necessary to enable a process of continual improvement.

Other upstream measurements could include the following:

- Risk assessments conducted
- Attendance at safety demonstrations and DVD presentations
- Number of contractor site inspections
- Contractor safety meetings held
- JSAs reviewed and updated
- Safety suggestions received and implemented

- Ergonomic studies completed
- Number of safe behaviors reported
- Number of Safety Representatives reappointed and trained
- Department audit scores
- Change management projects completed
- Safety commendations issued
- Safety culture survey results and scores
- Permit Issuers and Receivers retrained
- Safety competitions held
- Planned Job Observations (PJOs) conducted
- Incident and accident recall sessions conducted, etc.

SAFETY SYSTEM DEVELOPMENT AND IMPLEMENTATION

Since a fully fledged safety system will be implemented over a time period, the degree of implementation should be monitored on a monthly basis. This progress can also be quantified and ranked on a 1–5 scale for each element, and a total score obtained for the entire system's implementation progress. This measurement of the safety system development can be kept in the form of a table as in Figure 15.4. There are five criteria on this table which are each scored on a 1–5 scale:

1. The writing of the standard is scored 1–5 for each element of the safety system written.
2. The standard's approval and signature by the approval authority ranks on a 1–5 scale.
3. If the standard has a checklist or supporting document, such as a training program or similar, this is rated on a 1–5 scale.
4. Training of employees and committees on the contents of the standard also rates 1–5. If, for example, only 50% of employees have been trained on the standard, then a score of 2.5 would be allocated.
5. Depending on the degree of implementation of the standard within the company, a ranking of 1–5 is allocated.

Item	Name of element (Element number and title)	1	2	3	4	5
1	Safety system standard written (Rank 1–5)					
2	Standard approved and signed off by the leadership (Rank 1–5)					
3	Checklist or support document (Rank 1–5)					
4	Training of employees in the content of the standard (Rank 1–5)					
5	Degree of implementation of the standard (Rank 1–5)					
	Total score					

FIGURE 15.4   Measuring the safety system development and degree of implementation.

The total maximum scores, and actual score achieved can be expressed as a percentage of the system implementation, which is a figure that can be included in the action plan. For example:

- Year 1—50% achievement
- Year 2—60% achievement
- Year 3—70% achievement
- Year 4—80% achievement
- Year 5—90% achievement

COMBINATION

Ideally a combination of upstream and downstream indices should be used to measure the entire spectrum of safety effort and experience performance. No one single measure should be used, but rather a number of measures so that a complete snapshot of the safety at the organization is presented.

16 Case Study

THE ORGANIZATION

The organization is a country-wide generator, transmitter, and distributor of electricity, employing some 30,000 employees and 25,000 contractors. More than 112 Auditable Units were identified across the country, which included power plants, distribution depots, warehouses, and medical and training centers.

SITUATION AS IT WAS

There was no formalized Safety Management System (SMS) in place. Safety was primarily the responsibility of the safety department which was both centralized and decentralized. Safety fell under the same banner as security, and the dividing lines were not clear. What was clear, was that all levels of management had little to do with safety, which was left over to the Safety Coordinators, who were seen as safety policemen, more than advisors.

Most safety staff had no formal safety qualifications, but were mostly employees who had Bachelor's degrees in mechanical or electrical fields. Some had completed the British NEBOSH (National Examination Board for Occupational Safety and Health) basic examination. Job descriptions for the safety personnel were nondescript, inadequate, and did not describe safety functions at all.

Numerous fatalities were experienced every year, many as a result of road accidents, and many included contractors' employees. During analysis, it was found that the reported injury rate was 10 times lower than the U.S. national average for that industry, but that the fatality rate was 8 times higher.

Since the injury rate was the only gauge of safety performance, there was a culture of under reporting and hiding injuries. Safety services were charged out to the different departments, and since they were balancing their budgets, the departments never requested safety services.

The Safety Coordinator did inspections at facilities on their own and send a report to the manager, which was never heard of again.

STRATEGY

The objective was to set up a structured safety management system throughout the company. A safety consultant was contracted to facilitate the project, which started with a presentation of a strategy to the executives. The proposal was accepted and the timeline estimated at four to five years, to fully implement a world's best practice system.

An executive director took ownership of the project and met with the consultant and corporate safety team weekly. The executive safety committee (EXCO), chaired by the CEO, was formed, as well as four levels of regional safety committees across the organization. A safety and health policy was drafted and signed by all 12 executives.

A five-year detailed action plan was compiled, approved, and set as the target for the organization. Targets were set to achieve 91% compliance to the standards after five years, and to develop and implement 14 standards each year. After much discussion, injury rates were replaced by proactive safety performance indicators (SPIs), which each departmental manager had to report on during the weekly telephonic conference. No longer did the safety manager report alone, now managers had to account for safety.

One of the objectives was to train 600 managers and 600 supervisors per year, in the working of the safety system and their responsibilities in safety. More than 3,000 Safety Representatives were appointed and trained on a special one-day program throughout the country. This led to an immediate return of 3,000 safety inspections being done each month. The training was to be repeated every year. Booklets containing a summary of the 75 elements of the safety system were printed and distributed to all levels in the company.

Eight safety consultants, who were well experienced in safety system implementation, were contracted and decentralized to the four main geographic areas. These areas also had operating safety committees and local safety departments.

The occupational health program was expanded to include all operational areas, and hearing acuity tests were extended to all regions. A start was made on the industrial hygiene program.

SAFETY STRUCTURE

Despite efforts, the safety structure was not changed. Having safety fall under security was not an ideal situation. All safety personnel attended training workshops on the safety system, up to internal accredited auditor level. This entailed more than 100 hours of classroom and on site instruction. New job descriptions were approved, and new performance indicators were agreed to.

SAFETY MANAGER

The Safety Manager who was appointed as the project leader, turned out to be a key player in the success of the launch and implementation of the system. He was able to communicate at executive level and managed to get things done, despite red tape and resistance to change. He facilitated a number of initiatives and kept the momentum of the ground swell. He motivated the safety departments and earned the respect of managers and employees with his pleasant manner and ability to achieve objectives. Such a safety disciple is vital to the success of a safety management system implementation project.

BASELINE AUDITS

Baseline audits were conducted after a year by three-man teams. Each audit took 2 days. One day was spent doing the physical inspection and the next, the control documentation verification. More than 100 Auditable Units were audited. Each auditor allocated their own score which was averaged to give the final percentage.

The results shocked some departments (who had made no effort) and pleased others, who had started the implementation of the safety system. The audits caused a ripple throughout the organization as the results were publicized company wide. The first year target was to achieve an audit score of 50%, the next year 60%, the third year 70%, and at the end of five years 95%. Scores were tabled and discussed at the EXCO meetings.

Some directors, realizing that the system entailed hard work and effort, kicked against the system and called the system ineffective. Reviews of other safety systems in use took place, but eventually, the CEO decided that the organization stay with the original system.

NEAR MISS INCIDENT SUCCESS

Employees were encouraged to report near miss incidents. A small, pocket size reporting booklet was distributed to all, and soon the organization was flooded with reports. Employees, and even line managers, became involved in the process of safety reporting.

FIRST ELEMENTS

One of the first elements implemented was Business Order, as visits had confirmed that the physical work areas were well below standard. Soon managers were visiting the workshops and helping clean up, demarcating walkways, and improving stacking areas. These basic elements created a ground swell of enthusiasm towards safety that formed the foundation of the system. All were now becoming involved in safety.

Despite criticism of starting with the basics of housekeeping and workplace neatness, it built a good platform on which to launch more involved standards such as work permits, critical tasks, etc. As Frank E. Bird said, "If you clean up the workplace, you clean up the thought processes of the people in that workplace."

The approach led to the physical transformation of the workplaces. Floors were cleared of debris, walkways were demarcated, windows cleaned, and lights repaired. Old safety posters were removed from walls and the environment brought up to a standard that is expected of a world leader in safety.

NEWSLETTER

A monthly safety newsletter was sent out to all employees and safety was publicized on the intranet, on notice boards, and in the company magazine. Safety was now on the map within the company.

LOGO AND IDENTITY

A safety system logo was adopted and the system now had an identity. A committee representing the country-wide operations, met monthly to review and approve safety standards and to report on regional progress in safety.

TRAINING OF STAFF

Discussions were held with two universities to start a Bachelor's and Master's degree program for further education of current and future safety personnel. A selected group of Safety Coordinators were sent to the United States for on-site training in auditing.

REGIONAL COMMITTEES

Regional committees met regularly to monitor system implementation progress, and the safety staff and consultants attended these forums acting as guides and advisors on the safety system. They now gave a service to the client. Previously, they would do an unaccompanied inspection and send a report, which was never read, to the manager, whom they had never met. As each standard was released, a meeting was held where the implementation of the requirements of the standard was planned.

CONTRACTORS

Recording of injuries and fatalities now included all contractors, who were also made a part of the safety system, and who had to meet the system standards on all their sites. Contractors were now compelled to abide by the requirements of the safety system, and were now regarded as company employees. For the first time, a contractor's site was shut down because of safety hazards found during one inspection. This caused an uproar throughout the organization, but the safety inspector stood his ground and the organization supported his decision. The problems were rectified and the contractor was permitted to continue work. The ice had been broken.

REGIONAL SEMINARS

Each region's annual conference was dominated by presentations on the safety system and knowledge about the system spread. To reward their efforts, Safety Representatives were invited to attend the conferences. A Safety Representative of the Year competition was planned for the future.

MANAGEMENT SETS THE EXAMPLE

Managers now wore correct work clothes and appropriate PPE when they ventured into the workplace. They started to follow the example being set by the CEO. During audits, they accompanied the auditors on the inspections and presided at the control documentation review conferences, and audit close-out meetings. The system of paying for safety services was dropped, and managers now requested safety staff to visit them to help them improve their safety system, especially prior to audits. There was now a sense of urgency to meet the targeted percentage score.

The CEO took the lead and visited workplaces regularly and practiced visible felt leadership, setting an example to his managers who were mostly office bound. He appointed another senior executive as a *Safety and Health Champion*, and another safety initiative was formed to support the safety drive.

SOME HIGHLIGHTS

The safety system implementation brought about a major change in the safety culture of the organization. Risk assessments were carried out on a daily basis at worksites. Managers, contractors, and employees were now aware of the safety system and participated in meetings, inspections, and discussions. Managers took ownership of safety in their areas of responsibility. Safety staff acted as advisors and consultants to managers, employees, and contractors. The safety system had an identity and the safety website was a major source of information. Targets were set to write all 75 element standards and monthly reports on system development progress were submitted. Safety notice boards appeared in all departments and the safety policy was printed, framed, and circulated to all departments. Every meeting started with the topic of safety. The organization's insurance company showed an interest in the safety system development, and the risk management department participated in the central safety committee.

Safety system standards were written, approved by the central committee, and cascaded down to the employees via discussions and presentations at regional level meetings.

A comprehensive road safety program was also launched. Some major contractors soon turned to the organization for advice on the safety system, which was proving to be world's best practice.

Other initiatives included the following:

- Regular evacuation drills
- Revamp of the fire equipment maintenance system
- Lockout system revised and improved
- Work permits system revamped
- Using safety personnel as advisors, rather than enforcers
- Office building inspections and ratings
- SMS logo on safety promotional gifts
- Electrical arc flash risk assessments and training started
- PPE selection committee re-elected
- Incident and accident recall sessions started, etc.

ONGOING

The process is still ongoing, with each day bringing new safety actions and innovations to the table. Slowly but surely, safety is becoming a part of the daily routine at the organization, thanks to the safety management system. Progress has been made, but the final objective has still to be reached.

References

American National Standards Institute (ANSI). 2015a. http://webstore.ansi.org/RecordDetail. aspx?sku=ANSI%2fAIHA%2fASSE+Z10-2012.

American National Standards Institute, Inc. (ANSI). Z16.1. Measuring and Recording Injury Experience.

American National Standards Institute (ANSI). Guidelines: ANSI/ASSE Z590.2-2003. *Criteria for Establishing the Scope and Functions of the Professional Safety Positions.*

Bird, F. E. Jr. and Germain, G. L. 1992. *Practical Loss Control Leadership* (2nd ed.). Loganville, GA: International Loss Control Institute (ILCI).

British Safety Council (BSC). 2015. *Five Star Occupational Health and Safety Audit Specification.* London: British Safety Council, p. 3.

British Safety Council (BSC). 2016. https://www.britsafe.org/audit-and-consultancy/five-star-occupational-health-and-safety-audit#sthash.4KU619yA.dpuf.

British Standards Institute (BSI). 2015. BSI-OHSAS 18001. *Occupational Health and Safety Management Systems.* http://www.bsigroup.com/en-ZA/BS-OHSAS-18001-Occupational-Health-and-Safety/ (Permission to reproduce extracts from BSI-OHSAS 18001 is granted by BSI).

British Standards Institute (BSI). 2016. http://www.bsigroup.com/LocalFiles/nl-nl/bs-ohsas-18001/resources/BSI-BS-OHSAS%2018001-Implementing-Guide-EN-NL.pdf.

DNV GL. 2015. https://www.dnvgl.com/services/isrs-for-the-health-of-your-business-2458.

European Agency for Safety and Health at Work. 2012. (EU-OSHA) *Management Leadership in Occupational Safety and Health – A Practical Guide.* pp. 10–14.

Fortune.com. http://fortune.com/2015/07/13/bp-billions-compensation-claims/?xid=yahoo_fortune.

Heinrich, H. W. 1959. *Industrial Accident Prevention* (4th ed.). New York: McGraw-Hill Book Company, p. 21.

International Labor Organization (ILO). 2016. http://www.ilo.org/global/about-the-ilo/media-centre/press-releases/WCMS_007969/lang--en/index.htm.

International Labor Organization (ILO). 2015. http://www.ilo.org/global/about-the-ilo/lang--en/index.htm.

International Labor Organization (ILO). 2001. *Guidelines on Occupational Safety and Health Management Systems,* ILO-OSH 2001. p. 2.

International Organization for Standardization (ISO). 2009. *ISO 31000: 2009 Risk Management– Principles and Guidelines.* pp. 1–2.

International Organization for Standardization (ISO). 2015a. *ISO 45001 Briefing Notes.* p. 2.

International Organization for Standardization (ISO). 2015b. *ISO 9001: 2015 Quality Management Principles.* pp. 1–7.

International Organization for Standardization (ISO). 2016a. http://www.iso.org/iso/home/about.htm.

International Organization for Standardization (ISO). 2016b. *ISO 14001 Key Benefits.* p. 2.

McKinnon, Ron C. 2000. *The Cause, Effect, and Control of Accidental Loss, With Accident Investigation Kit. (CECAL).* Boca Raton, FL: CRC Press (Preface–Theory).

McKinnon, Ron C. 2007. *Changing Safety's Paradigms.* Lanham, MD: Government Institutes, p. 135.

National Fire Protection Association (NFPA). 2015. http://www.nfpa.org/codes-and-standards/document-information-pages?mode=code&code=101. Reproduced with permission fromhttp://www.nfpa.org/codes-andstandards/document-information-pages?mode=code&code=101, Copyright © 2015, National Fire Protection Association.

National Occupational Safety Association (NOSA). 1995. *The NOSA 5 Star System.* (Vol. HB0.0050E.) Pretoria: National Occupational Safety Association.

National Safety Council (NSC). 2013. *Injury Facts.* (2013 Ed.). Itasca, IL: Author. Library of Congress Catalog Number: 99-74142.

Occupational Safety and Health Administration (OSHA). 2015a. OSHA 1910.27 *Fixed ladders.*

Occupational Safety and Health Administration (OSHA). 2015b. https://www.osha.gov/Top_Ten_Standards.html.

Occupational Safety and Health Administration (OSHA). 2015c. https://www.osha.gov/SLTC/bloodbornepathogens.

Occupational Safety and Health Administration (OSHA). 2015d. https://www.osha.gov/recordkeeping.

Occupational Safety and Health Administration (OSHA). 2016. https://www.osha.gov/dcsp/vpp/index.html.

Occupational Safety and Health Administration (OSHA). OSHA 1910.179 *Overhead and Gantry Cranes.*

Occupational Safety and Health Administration (OSHA). 1910.184 *Slings.*

Ohio State University (OSU). Columbus, OH. (Environmental Health and Safety) *Heat and Cold Stress Program.* http://ehs.osu.edu/FileStore/Occ%20Health%20&%20Safety/Heat%20and%20Cold%20Stress%20Program.pdf.

Index

Note: Page numbers followed by f and t refer to figures and tables, respectively.